Signals and Communication Technology

This series is devoted to fundamentals and applications of modern methods of signal processing and cutting-edge communication technologies. The main topics are information and signal theory, acoustical signal processing, image processing and multimedia systems, mobile and wireless communications, and computer and communication networks. Volumes in the series address researchers in academia and industrial R&D departments. The series is application-oriented. The level of presentation of each individual volume, however, depends on the subject and can range from practical to scientific.

Indexing: All books in "Signals and Communication Technology" are indexed by Scopus and zbMATH

For general information about this book series, comments or suggestions, please contact Mary James at mary.james@springer.com or Ramesh Nath Premnath at ramesh.premnath@springer.com.

Chua-Chin Wang • Arumugam Nallanathan
Editors

6th International Conference on Signal Processing and Information Communications

ICSPIC 2023

Springer

Editors
Chua-Chin Wang
National Sun Yat-sen University
Kaohsiung, Taiwan

Arumugam Nallanathan
School of EE and Computer Science
Queen Mary University of London
London, UK

ISSN 1860-4862 ISSN 1860-4870 (electronic)
Signals and Communication Technology
ISBN 978-3-031-43780-9 ISBN 978-3-031-43781-6 (eBook)
https://doi.org/10.1007/978-3-031-43781-6

This Springer imprint is published by the registered company Springer Nature Switzerland AG
The registered company address is: Gewerbestrasse 11, 6330 Cham, Switzerland

Paper in this product is recyclable.

Preface

This volume of proceedings gathers the papers presented at the 2023 International Conference on Signal Processing and Information Communications (ICSPIC 2023) held virtually from February 25 to 27, 2023. Due to the long-term impact of COVID-19, the participants were unable to meet in Singapore as planned. Though the authors and speakers couldn't communicate face to face, the passion for involvement is active.

ICSPIC was held successfully in Seoul, Phuket, and Bali and was held virtually in 2021 and 2022. ICSPIC 2023 successfully hosts a global gathering of theorists and experts in advanced characterization techniques in the highly complex field of signal processing and information communications, which brings researchers, practitioners, and scientists into the discussion of the latest methods, research developments, and future opportunities. The successful hosting of ICSPIC 2023 contributed to the development of signal processing and information communication.

The organizing committee of ICSPIC 2023 used Zoom as the supportive platform to allow all the participants around the world to gather online at the same time. The test session was arranged to make sure authors will not be affected by the technical issues during the presentations. There are three keynote speeches presented at ICSPIC 2023. It gives all participants an opportunity to have an unforgettable discussion in the field of signal processing and information communication.

This volume of proceedings contains nine papers. In the proceedings, readers can learn the most cutting-edge knowledge about signal processing and information communication from all around the world. And each paper was presented by the author correspondingly. Topics include digital signal and image processing, modern information technology and management. Each selected paper has undergone a double-blinded peer review.

I am very grateful for the participation of all authors who shared the present research in signal processing and information communication. In addition, our sincere gratitude goes to the program chairs, publicity chairs, international reviewers, and conference secretariat who are dedicated to making the conference run smoothly and properly and ensuring the quality of the proceedings. Without their active action, the conference couldn't be held successfully. Finally, ICSPIC welcomes all of you who are interested in helping us to make it a glorious future.

Chua-Chin Wang

Department of Electrical Engineering/
Institute of Undersea Technology,
National Sun-Yiet Sun University,
Kaohsiung, Taiwan

Conference Committees

Conference General Chair
Chua-Chin Wang, National Sun Yat-Sen University, Taiwan

Conference Co-chairs
Chip Hong Chang, Nanyang Technological University, Singapore

Program Chairs
Kai-Kit Wong, University College London, UK
Arumugam Nallanathan, Queen Mary University of London, UK

Program Co-chairs
Wee Peng Tay, Nanyang Technological University, Singapore
Eng Siong Chng, Nanyang Technological University, Singapore

Publicity Chair
Adao Silva, University of Aveiro, Portugal

Technical Program Committees
Waleed H. Abdulla, The University of Auckland, New Zealand
Dimitris Ampeliotis, Ionian University, Greece
Muhammad Naveed Anwar, Northumbria University, UK
Guillermo E. Atkin, Illinois Institute of Technology, USA
Paul Bogdan, University of Southern California, USA
Lin Cai, Illinois Institute of Technology, USA
Aleksandr Cariow, West Pomeranian University of Technology, Poland
Stefano Cirillo, University of Salerno, Italy
Joydev Ghosh, National Research Tomsk Polytechnic, Russia
Sean Mc Grath, University of Limerick, Ireland
Artyom Grigoryan, The University of Texas at San Antonio, USA

Pietro Guccione, Politecnico di Bari, Italy
Sami Ahmed Haider, University of Worcester, UK
Lim Tiong Hoo, Universiti Teknologi Brunei, Brunei
Faheem A. Khan, University of Huddersfield, UK
Krzysztof Kulpa, Warsaw University of Technology, Poland
Grigorios L. Kyriakopoulos, National Technical University of Athens (NTUA), Greece
Tsung-Jung Liu, National Chung Hsing University, Taiwan
Derya Malak, University of Minnesota, UK
Lyudmila Mihaylova, The University of Sheffield, UK
Hasan S. Mir, American University of Sharjah, UAE
Konstantinos Nikitopoulos, University of Surrey, UK
Eduardo Manuel Godinho Rodrigues, Universidade de Aveiro, Portugal
Cristina Rottondi, Politecnico di Torino, Italy
Rachid Sabre, University of Burgundy, France
Zeljen Trpovski, University of Novi Sad, Serbia
Oleksii K. Tyshchenko, University of Ostrava, Czech Republic
Bo Wei, Northumbria University, UK
Zeynep Yucel, Okayama University, Japan
Chi Zhou, Illinois Institute of Technology, USA
Ladislav Polak, Brno University of Technology, Czech Republic

Contents

Chapter 1
Automatic Recognition of Wild Animals for Road Accident Prevention Using Deep Learning with Yolov4

Papa Assane Diop, Amadou Dahirou Gueye, and Malal Deme

Abstract In this paper, we address the problem of wildlife recognition for road accident prevention, where a rate of 63.17% per year of road accidents is noted in Senegal. Given that the movement of animals is unpredictable in spite of road signs, this constitutes a handicap in preventing wild animals that may cause an accident. The solution proposed in this paper allows real-time detection of wild animals crossing roads, especially in non-built-up areas. It is based on computer vision using deep learning with the Yolov4 approach which allowed the categorisation of the three types of wild animals chosen in this paper: cows, donkeys and goats. The choice of these three types of animals is justified by the fact that most wildlife-related road accidents are caused by these types of animals. To achieve this, we first collected a set of images of the three types of animals. These collected images are sorted before the model is created. The evaluation of the model was carried out using test images and also videos taken on the Niague road, more precisely in the suburbs of the Senegalese capital, with accuracy rates of 94, 32%, 98.85 and 99.96%, respectively, for cows, goats and donkeys.

Keywords Obstacle · Road · Deep learning · Detection · Yolov4 · Recognition · Wild

1.1 Introduction

Developing countries are facing an increase in the number of vehicles in regional capitals. There is a high density of vehicle movement from region to region. In Senegal, road traffic accidents account for 63.17% of all accidents, which means that [1] of all accidents, which means that the majority of accidents are related to road insecurity. These traffic accidents occur outside built-up areas, i.e. between regions of the same country or between different countries. These accidents in nonurban

P. A. Diop (✉) · A. D. Gueye · M. Deme
TIC4DEV TEAM, Alioune DIOP University of Bambey, Bambey, Senegal
e-mail: papaassane.diop1@uadb.edu.sn

C.-C. Wang, A. Nallanathan (eds.), *6th International Conference on Signal Processing and Information Communications*, Signals and Communication Technology, https://doi.org/10.1007/978-3-031-43781-6_1

areas are most often caused by wildlife species. The authors in [2] conducted research to assess the impact of wildlife-vehicle collisions along the Dakar-Bamako corridor on animal populations in the Niokolo Koba National Park.

With all these problems related to roads, video surveillance is a necessary means of ensuring road safety. Video surveillance, commonly known as video protection, is made up of cameras and everything useful to record and exploit the images in order to detect abnormal events [3]. The main objective of processing a digital image is to extract information and improve its visual quality in order to make it more interpretable by a human analyst or an autonomous machine perception.

The use of video surveillance allows us to use computer vision which includes many object detection models. In our previous work [4], we proposed a lateral road obstacle detection model based on machine learning to contribute to road safety in areas outside built-up areas. Today, with the use of deep neural networks in computer vision, deep learning is taking over machine learning in terms of video surveillance. In this paper, we have chosen YOLOV4 for the detection of three types of animals in the wildlife which are cows, donkeys and goats. The rest of the paper will be structured as follows: in Sect. 1.2, we present related work on obstacle detection systems in the field of road video surveillance. In Sect. 1.3, we propose an approach based on the YOLOv4 detection model. In Sect. 1.4, we present the training and testing results of the model as well as the performance metrics. In Sect. 1.5, we conclude with a conclusion and perspectives.

1.2 Related works

In this section, we will look at work on animal detection and monitoring using deep learning, specifically yolo (You Only Look Once).

1.2.1 Object Detection Models Based on Deep Learning

The past decade, object detection models based on deep learning have gained great importance in the research field. In [5–7], there is a good overview of the state of the art of object detection models based on deep learning. For example, in [5], the author shows us that with the advancement of artificial intelligence, neural networks such as convolutional neural networks (CNNs) were often used in image processing. Later, CNN models face many problems in execution, performance, deployment, etc. In [8], another deep learning network, namely, Faster Regional Convolution Neural Network (Faster R-CNN) for object detection and tracking, is discussed. In the literature, we note other types of algorithms such as SSD [9] and F-CNN [10]. In this paper, we choose the deep learning detection algorithm Yolov4 which is much faster and more efficient in terms of detection [11]. These algorithms have often been used in video surveillance for object detection.

1.2.2 Roadside Video Surveillance of Wild Animals

In recent years, video surveillance has been the subject of much research using deep learning. Deep learning also gives rise to detection techniques such as YOLO. A presentation of the state of the art is available in [5, 9, 12].

For example, in [9], Haomin and He proposed a study on YOLO object detection algorithm for road scenes based on computer vision. They made a study on Yolo detection algorithms at the road level based on computer vision. The authors in [12] made an in-depth study on the progress of road object detection optimisation, which is an important part of detection and also the evaluation of detection models.

As pointed out by [13], the Yolo detection models withstand conditions such as night, rain and snow to provide fast and reliable detection. In [14], the authors presented a publication on the detection of wild animals in the forest and their use to monitor their movement. In the survey of research on detection models in road safety, we did not find any work on the Yolov4 detection model to warn the wildlife crossing roads. Thus, our paper is based on the Yolov4 approach to perform automatic wildlife detection applied on roads for accident prevention. In the following, we will present the detection approach based on YOLOV4.

1.3 Detection Approach Based on Yolov4

1.3.1 Architecture of Yolov4

In [15], the Yolov4 architecture is made up of different parts. The input comes first, and this is essentially what we have as our set of training images that will be passed to the network – they are processed in batches in parallel by the GPU. Then comes the backbone and the neck which does the feature extraction and aggregation. The sensing neck and sensing head can be referred to as an object detector assembly (Fig. 1.1).

YOLOv4 explores different backbones and data augmentation methods:

- Backbone network
- Neck
- PANet (Path Aggregation Network)
- Head

The head is the main function; it is to locate the selection frames and perform the classification.

The coordinates of the selection frame (*x*, *y*, height and width) and the scores are detected. Here, the *x* and *y* coordinates are the centre of the *b*-box expressed relative to the grid cell boundary. The width and height are predicted relative to the whole image.

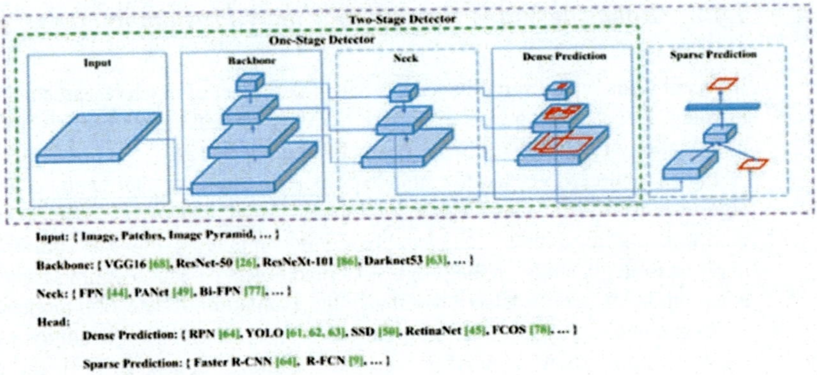

Fig. 1.1 Object detector

$$b_x = \sigma(t_x) + C_x \tag{1.1}$$

$$b_y = \sigma(t_y) + C_y \tag{1.2}$$

$$b_w = p_w e^{tw} \tag{1.3}$$

$$b_h = p_h e^{th} \tag{1.4}$$

1.3.2 Construction of Our Dataset

Our data represents a collection of images of three types of wild animal species: cows, goats and donkeys. These data were acquired through Google search sites, on a farm in Senegal, specifically in Niague, which raises cows. After the collection, we renamed the images using python code to make the renaming faster. Before renaming, we did a very essential step which is to remove the irrelevant images. After that, a problem arises, because the images acquired through the websites and the images taken through a camera on a farm were not the same size, so we have to do a resizing so that the size of all the images conforms to 671 × 480. Finally, we labelled the images. We used labelImg which is an open-source image annotation tool. We have 1000 images for each type of animal, making 3000 images in total (Fig. 1.2).

Fig. 1.2 Donkey, goat and cow

1.4 Experimentation and Validation

In this section, we will show the details of training our model to detect three (03) classes (donkey, goat and cow). Then, we present the performance measures of our model.

1.4.1 Setting Up the Experiments

Implementation Details For the training of the YOLO model, we based ourselves on the Darknet framework which contains all the necessary files [16]. The training phase of YOLO requires a lot of time, which is why we use the transfer learning method. This method consists of dividing the training phase between deep artificial neural networks, which results in savings in machine resources and computing time. We need to use a pre-trained model of YOLO to do the transfer learning. Before starting the training, we need to make some settings to adapt it to our model. These modifications concern the number of classes, the number of iterations and the number of filters to be used at the layer level of the convolutional neural networks.

Training Environment The training phase of a YOLO model is rather heavy, and if you have a lot of images, you will need to have a machine with very powerful resources (GPUs, RAM) for the model to learn in a suitable time frame. This is why we use Google Colab Pro to train our data [17].

Splitting the Dataset (Training/Test) We will just split our dataset (1000 images per class) to have a training dataset (80%) and a test dataset (20%). So we will have the following:

- A training data set (80%)
- A test data set (20%)

1.4.2 Performance Measures of Our Model

Several indicators can be used to measure the performance of an object detection model. Each one has its own specificities, and it is often necessary to use several of them to have a complete view of the performance of a model. Most of these indicators depend on the parameters true positive (TP), false positive (FP), false negative (FN) and true negative (TN) [18].

- TP: These are the correctly predicted positive values, which mean that the actual class value is yes and the predicted class value is also yes.
- TN: These are the correctly predicted negative values, which means that the actual class value is no and the predicted class value is also no.
- FP: When the actual class is no and the predicted class is yes.
- FN: When the actual class is yes but the predicted class is no.

Now we will define the performance measurement indicators for the case of a YOLO model [18].

Accuracy (*P*) The accuracy is the number of objects correctly assigned to class i relative to the total number of objects predicted to belong to class *i*.

$$P = \frac{\mathrm{TP}}{\mathrm{TP} + \mathrm{FP}} \tag{1.5}$$

Recall (*R*) Recall is the number of objects correctly assigned to class i out of the total number of objects belonging to class i.

$$R = \frac{\mathrm{TP}}{\mathrm{TP} + \mathrm{FN}} \tag{1.6}$$

F_1-Score (F_1) Although useful, neither precision nor recall can fully evaluate a model. The F_1-Score provides a good assessment of the performance of our model. The F_1-Score subtly combines precision and recall to make a good assessment of a model's performance.

$$F_1 = 2 * \frac{P * R}{P + R} \tag{1.7}$$

Intersection on Union (IoU) It indicates the overlap of the coordinates of the predicted bounding box with the ground truth box. A higher IoU indicates that the coordinates of the predicted bounding box closely resemble the coordinates of the ground truth box.

$$\mathrm{IoU} = \frac{\text{Area of overlap}}{\text{Area of union}} \tag{1.8}$$

Mean Average Precision The mAP is calculated by finding the average precision (AP) for each class, then averaging over the total number of classes. Interestingly, the average precision (AP) is not the average of the precision (P). The term AP has evolved over time. To simplify, it can be said to be the area under the precision-recall curve. The mAP incorporates the trade-off between precision and recall and considers both false positives (FP) and false negatives (FN). This property makes mAP a suitable metric for most detection applications [19].

$$\mathrm{mAP} = \frac{1}{n}\sum_{i=0}^{n} \mathrm{AP}_i \tag{1.9}$$

Loss Function This is the sum of the errors made for each example in training sets. The main objective of a learning model is to minimise the value of the loss function with respect to the model parameters by modifying the values of the weight vector using different optimisation methods, such as back-propagation in neural networks. AP_i

1.4.3 Results and Analysis

After training, a graph is generated. The graph shows us the evolution of the average accuracy (mAP) of the model and the loss function as a function of the iterations (Fig. 1.3).

This graph shows that after 1000 iterations mAP = 72% then at 1200 iterations mAP = 98% then at 2500 iterations mAP = 99%, and in all remaining iterations, mAP is equal to about 98%. We also see that the loss function keeps decreasing until the end of the training to reach 0.466. This graph shows us the results in a global way, while we have three (3) classes. The following figure will give us in detail the results obtained (Fig. 1.4).

For donkeys, we have an accuracy of 99.96% with the number of true positives (TP) =156 and the number of false positives (FP) =30.

For the cows, we have an accuracy of 94.32% with TP =308 and FP =24.

For goats, we have an accuracy of 98.85% with TP =274 and FP =18.

Averaging the accuracies for our three classes, we have 97.71%. This shows that the detection model is acceptable.

1.4.4 Test

Testing on Images To do the detection on an image, we use a python script that takes images as input and makes a prediction with our model (Fig. 1.5).

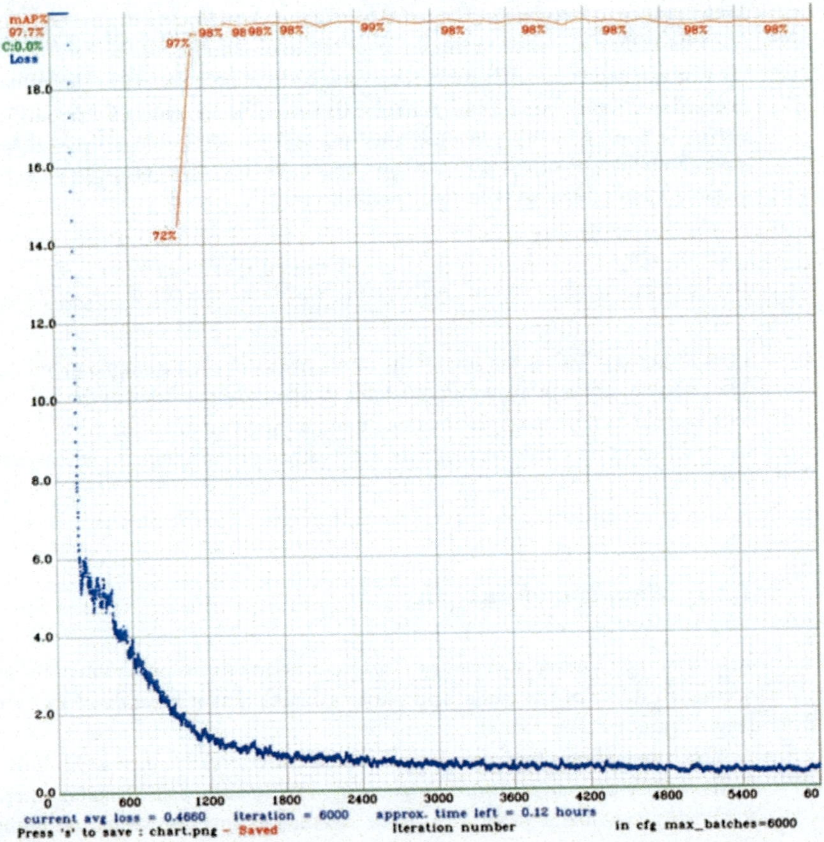

Fig. 1.3 Curve representing the evolution of the mAP and the loss function as a function of iterations

Testing on a Video To do the detection on a video, we use a python script that takes a video as input and makes a prediction with our model. This video was taken in real time on the Niague road located in Keur Massar, Senegal, as the cows are returning to the Niague farm after a day's walk (Fig. 1.6).

1.5 Conclusion and Outlook

Wild animals are increasingly unpredictable obstacles. Related work related to road obstacle detection has been presented to propose a Yolov4-based approach to detect wild animals such as cows, goats and donkeys crossing roads especially in nonurban areas. With this approach, a performance study of the model is done to validate our

```
 (next mAP calculation at 6080 iterations)
 Last accuracy mAP@0.50 = 97.73 %, best = 98.53 %
 6000: 0.253049, 0.466005 avg loss, 0.000010 rate, 5.946519 seconds, 384000 images, 0.125020 hours left
Resizing to initial size: 416 x 416  try to allocate additional workspace_size = 52.43 MB
 CUDA allocate done!

 calculation mAP (mean average precision)...
 Detection layer: 139 - type = 28
 Detection layer: 150 - type = 28
 Detection layer: 161 - type = 28
512
 detections_count = 982, unique_truth_count = 761
class_id = 0, name = ane, ap = 99.96%        (TP = 156, FP = 1)
class_id = 1, name = vache, ap = 94.32%      (TP = 308, FP = 24)
class_id = 2, name = chevre, ap = 98.85%     (TP = 274, FP = 18)

 for conf_thresh = 0.25, precision = 0.94, recall = 0.97, F1-score = 0.96
 for conf_thresh = 0.25, TP = 738, FP = 43, FN = 23, average IoU = 81.40 %

 IoU threshold = 50 %, used Area-Under-Curve for each unique Recall
 mean average precision (mAP@0.50) = 0.977066, or 97.71 %
Total Detection Time: 13 Seconds
```

Fig. 1.4 Detailed results of the training

Fig. 1.5 Detection from images of the three categories of animals: donkey, cow and goat

work. In the perspective of the work, we propose an integration of several types of animals and also an evaluation of the distance of obstacles. We also plan to integrate IOT devices for the deployment of our model in a vehicle.

Fig. 1.6 Detection of cow on a video obtained on the road of "Niague" Senegal

References

1. So that roads no longer kill in Senegal, https://www.afro.who.int/fr/news/pour-que-les-routes-ne-tuent-plus-au-senegal. Accessed 19 Aug 2022
2. S.M. Sarr, M. Gueye, A. Aziz, Impacts des collisions avec les véhicules le long du corridor routier Dakar-Bamako sur les populations fauniques du Parc National du Niokolo Koba, au Sénégal. 12 (2022)
3. I. Global, Role of CCTV cameras: Public, privacy and protection, https://www.ifsecglobal.com/video-surveillance/role-cctv-cameras-public-privacy-protection/. Accessed 03 Oct 2022
4. P.A. Diop, A.D. Gueye, A.K. Diop, Detection of lateral road obstacles based on the haar cascade classification method in video surveillance, in *Computer and Communication Engineering. CCCE 2022*, Communications in Computer and Information Science, ed. by F. Neri, K.L. Du, V.K. Varadarajan, S.B. Angel-Antonio, Z. Jiang, vol. 1630, (Springer, Cham, 2022). https://doi.org/10.1007/978-3-031-17422-3_3
5. Z. Li, J. Wang, An improved algorithm for deep learning YOLO network based on Xilinx ZYNQ FPGA, in *2020 International Conference on Culture-oriented Science & Technology (ICCST)*, (2020), pp. 447–451. https://doi.org/10.1109/ICCST50977.2020.00092
6. P.S. Kumar, V.P. Sakthivel, M. Raju, P.D. Sathya, A comprehensive review on deep learning algorithms and its applications, in *2021 Second International Conference on Electronics and Sustainable Communication Systems (ICESC)*, (2021), pp. 1378–1385. https://doi.org/10.1109/ICESC51422.2021.9532767
7. A.S. Abdullahi Madey, A. Yahyaoui, J. Rasheed, Object detection in video by detecting vehicles using machine learning and deep learning approaches, in *2021 International Conference on Forthcoming Networks and Sustainability in AIoT Era (FoNeS-AIoT)*, (2021), pp. 62–65. https://doi.org/10.1109/FoNeS-AIoT54873.2021.00023
8. K.B. Lee, H.S. Shin, An application of a deep learning algorithm for automatic detection of unexpected accidents under bad CCTV monitoring conditions in tunnels, in *2019 International Conference on Deep Learning and Machine Learning in Emerging Applications (Deep-ML)*, (2019), pp. 7–11. https://doi.org/10.1109/Deep-ML.2019.00010

9. H. He, Yolo target detection algorithm in road scene based on computer vision, in *2022 IEEE Asia-Pacific Conference on Image Processing, Electronics and Computers (IPEC)*, (2022), pp. 1111–1114. https://doi.org/10.1109/IPEC54454.2022.9777571
10. M. Maity, S. Banerjee, S. Sinha Chaudhuri, Faster R-CNN and YOLO based vehicle detection: A survey, in *2021 5th International Conference on Computing Methodologies and Communication (ICCMC)*, (2021), pp. 1442–1447. https://doi.org/10.1109/ICCMC51019.2021.9418274
11. Papers with Code – COCO test-dev benchmark (object detection), https://paperswithcode.com/sota/object-detection-on-coco. Accessed 03 Oct 2022
12. YOLO, You only look once – Real time object detection, https://www.geeksforgeeks.org/yolo-you-only-look-once-real-time-object-detection/. Accessed 19 Aug 2022
13. Y. Wendi, X. Yahang, Z. Xiaoyu, L. Jiaming, W. Tianchen, Risk assessment method combining trajectory prediction and lateral obstacle monitoring, in *2021 9th International Conference on Traffic and Logistic Engineering (ICTLE)*, (2021), pp. 1–5. https://doi.org/10.1109/ICTLE53360.2021.9525701
14. C. Zhu, T.H. Li, G. Li, Towards automatic wild animal detection in low quality camera-trap images using two-channeled perceiving residual pyramid networks, in *2017 IEEE International Conference on Computer Vision Workshops (ICCVW)*, (2017), pp. 2860–2864. https://doi.org/10.1109/ICCVW.2017.337
15. YOLOv4 model architecture, https://iq.opengenus.org/yolov4-model-architecture/. Accessed 26 Sept 2022
16. Darknet, Open source neural networks in C, https://pjreddie.com/darknet/, Accessed 03 Oct 2022
17. I. Ali, A. Khan, M. Waleed, A Google colab based online platform for rapid estimation of real blur in single-image blind deblurring, in *2020 12th International Conference on Electronics, Computers and Artificial Intelligence (ECAI)*, (2020), pp. 1–6. https://doi.org/10.1109/ECAI50035.2020.9223244
18. E. Solutions, Accuracy, precision, recall & F1 score: Interpretation of performance measures, https://blog.exsilio.com/all/accuracy-precision-recall-f1-score-interpretation-of-performance-measures/. Accessed 03 Oct 2022
19. Mean average precision (mAP) explained: Everything you need to know, https://www.v7labs.com/blog/mean-average-precision, https://www.v7labs.com/blog/mean-average-precision. Accessed 03 Oct 2022

Chapter 2
Continual Learning of Deep Learning for Indonesian Sentiment Analysis

Carlo Johan Nikanor, Hendri Murfi, Muhammad Adani Osmardifa, and Gianinna Ardaneswari

Abstract High-level social media usage makes this social media frequently used as one of the sources for sentiment analysis. Sentiment analysis is a field of study that analyzes people's opinions or evaluations of entities such as products and services. The Bidirectional Encoder Representation from Transformers (BERT) model is a deep learning architecture that achieves state-of-the-art performance for many natural language processing problems, including sentiment analysis. Several further developments have implemented continual learning on the deep learning model. By applying continual learning, the deep learning model continuously learns based on new data while retaining previously learned knowledge. In this paper, we analyze the performance of the BERT model for continual learning in some domains of Indonesian sentiment analysis. Then it will be compared with two standard deep learning models: fine-tuned embedding with CNN and fine-tuned embedding with LSTM. Our simulation shows the BERT model gives the best accuracy for the transfer of knowledge. However, the fine-tuned embedding with LSTM model is better for retain of knowledge. Moreover, our simulation shows that the order of the source domains affects the performance of BERT for both transfer of knowledge and retain of knowledge.

Keywords Sentiment analysis · Deep learning · BERT · Continual learning · Transfer of knowledge · Retain of knowledge

2.1 Introduction

Mobile applications have become an alternative solution for various needs during the pandemic. Bank Indonesia recorded an increase in transactions through e-commerce applications, namely, to 547 million transactions with a nominal value of IDR

C. J. Nikanor · H. Murfi (✉) · M. A. Osmardifa · G. Ardaneswari
Department of Mathematics, Universitas Indonesia, Depok, Indonesia
e-mail: hendri@ui.ac.id

C.-C. Wang, A. Nallanathan (eds.), *6th International Conference on Signal Processing and Information Communications*, Signals and Communication Technology, https://doi.org/10.1007/978-3-031-43781-6_2

88 trillion per the first quarter of 2021.[1] Shopee, Tokopedia, and Lazada are the top three e-commerce applications with the highest number of visitors per month.[2] This high level of use makes mobile applications collect many user opinions on their negative and positive features. Thus, we require machine learning for sentiment analysis on the opinion text data to provide information related to the advantages and disadvantages of the application or service of the mobile applications as a whole [1, 2].

From a machine learning point of view, sentiment analysis can be grouped as supervised learning because of sentiment labels [3, 4]. Deep learning is the primary machine learning method for unstructured data, such as text data. Deep learning extends the standard of machine learning by additional layers to extract a relevant representation of data [5]. The deep learning models widely used in sentiment analysis are convolutional neural networks (CNN) and long short-term memory (LSTM). Their efficiency and development have been mentioned in [6–8], including for Indonesian sentiment analysis [9]. The Bidirectional Encoder Representation from Transformers (BERT) model is another deep learning architecture that achieves state-of-the-art performance for many natural language processing problems [10]. BERT also improves the performance of standard deep learning for Indonesian sentiment analysis [11].

Continual learning, also known as lifelong learning or incremental learning, is the ability of a model to continuously learn based on new data while retaining previously learned knowledge [12]. The recommender systems on applications like Netflix and Amazon are well-known examples of continual learning. These applications instantly collect new labeled data as people interact with the applications. Continual learning algorithms have also succeeded in computer vision and clinical applications [13–15]. In practice, the main issue regarding continual learning is catastrophic forgetting, i.e., training a model with new information interferes with previously learned knowledge. This phenomenon typically leads to an abrupt performance decrease or, in the worst case, to the old knowledge being entirely overwritten by the new one.

In this paper, we analyze the performance of the BERT as a pretrained model of text data representation for continual deep learning in some domains of Indonesian sentiment analysis. Then it will be compared with two standard text data representations in deep learning: fine-tuned embedding with CNN and fine-tuned embedding with LSTM. Our simulation shows the BERT model gives the best accuracy for the transfer of knowledge. However, the fine-tuned embedding with LSTM model is better for retain of knowledge. Moreover, our simulation shows that the order of the source domains affects the performance of BERT for both transfer of knowledge and retain of knowledge.

---

[1] https://www.google.com/amp/s/ekbis.sindonews.com/newsread/472710/39/e-commerce-jadi-andalan-dongkrak-penjualan-di-masa-pandemi-162531223.

[2] https://www.webretailer.com/b/online-marketplaces-southeast-asia/.

The structure of this paper is as follows: in Sect. 2.2, we briefly explain methods. We describe the experiments in Sect. 2.3. Finally, a general conclusion about the results is presented in Sect. 2.4.

2.2 Methods

In this section, the methods used in this research will be explained. They are convolution neural networks (CNN), long short-term memory (LSTM), and Bidirectional Encoder Representations from Transformers (BERT).

2.2.1 *Convolutional Neural Network*

The convolutional neural network (CNN) is a deep learning model widely used in text classification. CNN uses filters to extract essential features from each region for text classification. During the process of word representation, the input will go through the convolution layer and the max-pooling layer (Fig. 2.1).

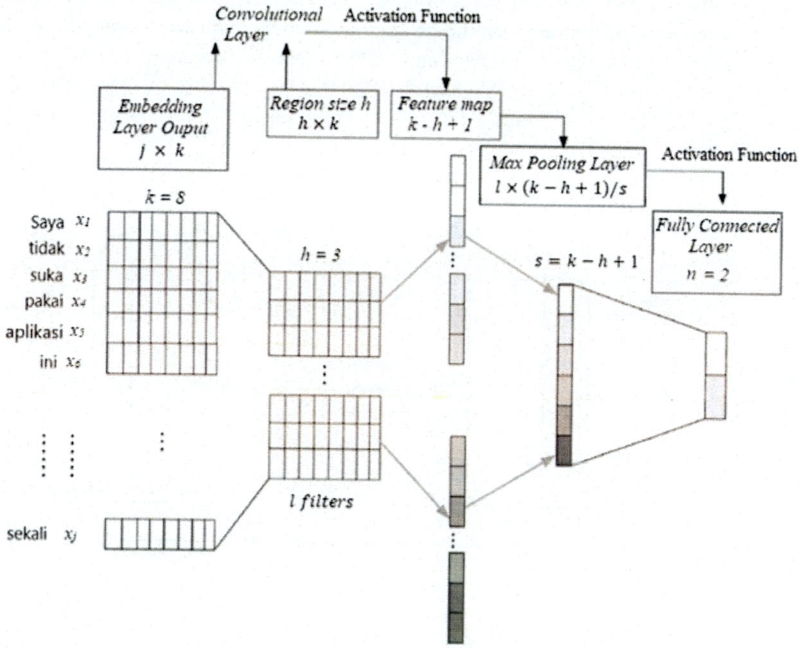

Fig. 2.1 CNN architecture

Convolution Layer In the convolution layer, the input will be processed by l filters W to find the essential features of each region with a specific region size. Suppose that the vector representation of the i-th word is denoted by x_i and the combination of the vectors of the words x_i to x_{i+h-1} is denoted by $X_{[i:i+h-1]}$. Eq. (2.1) calculates the feature vector $\boldsymbol{c} = [c_1, c_2, \ldots, c_{n-h+1}]$ for each filter, where j and k represent the rows and columns of the matrix, and f is a nonlinear activation function. The convolution layer's output is then used as the input for the max pooling layer.

$$c_i = f\left(\sum_{k=1}^{h}\sum_{j=1}^{d} X_{[i:i+h-1]k,j} \cdot W_{k,j}\right) \tag{2.1}$$

Max Pooling Layer The max pooling layer processes the output of the convolution layer by taking the essential features from each feature vector $\boldsymbol{c}$, that is $\hat{c} = \max\{\boldsymbol{c}\}$. The purpose of this layer is to reduce the dimension of the input, so the CNN will gradually learn to use less information with further iterations.

2.2.2 *Long Short-Term Memory*

Long short-term memory (LSTM) is a type of recurrent neural network (RNN) that aims to remember long-term information. The LSTM model has reasonable control over what information should be kept and removed at each training stage (time step) (Fig. 2.2). At the t-th time step, LSTM receives two input vectors which are the

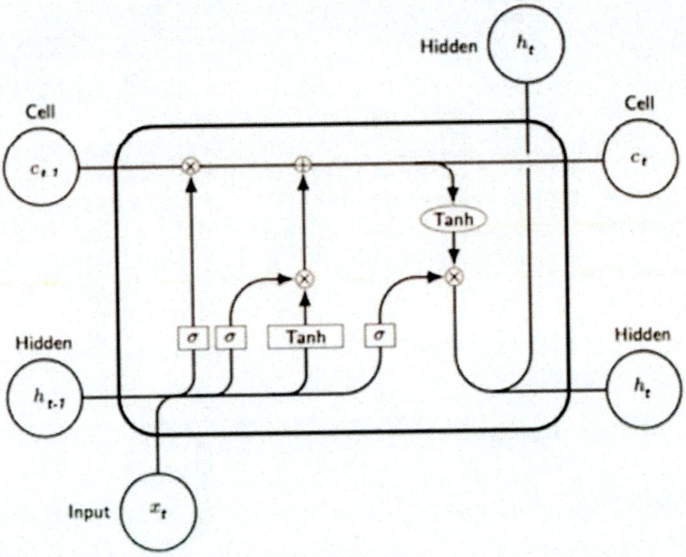

Fig. 2.2 LSTM architecture

vector representation of the t-th word in the sentence ($\boldsymbol{x}_t$) and the output vector of the previous hidden state ($\boldsymbol{h}_{t-1}$). The model will first determine what information should be removed from the cell state $\boldsymbol{C}_{t-1}$. This process is done at the forget gate ($\boldsymbol{f}_t$) shown in Eq. (2.2).

$$\boldsymbol{f}_t = \sigma\left(W_{fx}\boldsymbol{x}_t + W_{fh}\boldsymbol{h}_{t-1} + \boldsymbol{b}_f\right) \tag{2.2}$$

Next, the model will store selected information in the cell state $\boldsymbol{C}_t$. During this step, the model will also determine the value to be updated through the input gate ($\boldsymbol{i}_t$) as shown in Eq. (2.3) and the construction of a new vector that is the candidate cell state value ($\tilde{\boldsymbol{C}}_t$) in Eq. (2.4).

$$\boldsymbol{i}_t = \sigma(W_{ix}\boldsymbol{x}_t + W_{ih}\boldsymbol{h}_{t-1} + \boldsymbol{b}_i) \tag{2.3}$$

$$\tilde{\boldsymbol{C}}_t = \tanh(W_{cx}\boldsymbol{x}_t + W_{ch}\boldsymbol{h}_{t-1} + \boldsymbol{b}_c) \tag{2.4}$$

The cell state $\boldsymbol{C}_{t-1}$ is updated to a new cell state ($\boldsymbol{C}_t$) using the outputs of the forget gate, input gate, and candidate vector $\tilde{\boldsymbol{C}}_t$, as shown in Eq. (2.5).

$$\boldsymbol{C}_t = \boldsymbol{f}_t \cdot \boldsymbol{C}_{t-1} + \boldsymbol{i}_t \cdot \tilde{\boldsymbol{C}}_t \tag{2.5}$$

The last step for the model is to determine the output using the new cell state. This is done at the output gate ($\boldsymbol{o}_t$) shown in Eq. (2.6). The vector $\boldsymbol{o}_t$ will then be used with the cell state $\boldsymbol{C}_t$ to determine the hidden state $\boldsymbol{h}_t$ in Eq. (2.7). The vector $\boldsymbol{h}_t$ will be used in the next time step.

$$\boldsymbol{o}_t = \sigma(W_{ox}\boldsymbol{x}_t + W_{oh}\boldsymbol{h}_{t-1} + \boldsymbol{b}_o) \tag{2.6}$$

$$\boldsymbol{h}_t = \boldsymbol{o}_t * \tanh(\boldsymbol{C}_t) \tag{2.7}$$

where $W_{\{fx, fh, ix, ih, cx, ch, ox, oh\}}$ is a weight matrix and $\boldsymbol{b}_{\{f, i, c, o\}}$ is a bias vector [16].

2.2.3 Bidirectional Encoder Representation from Transformers

Bidirectional Encoder Representations from Transformers, commonly called BERT, is a trained language representation model developed by Devlin et al. [10]. Unlike the current language representation model, BERT does not use the traditional left-to-right or right-to-left language model. However, BERT is designed to train a bidirectional representation that simultaneously looks at each layer's left and right contexts. The main architecture of BERT is the transformer's encoder layers (Fig. 2.3).

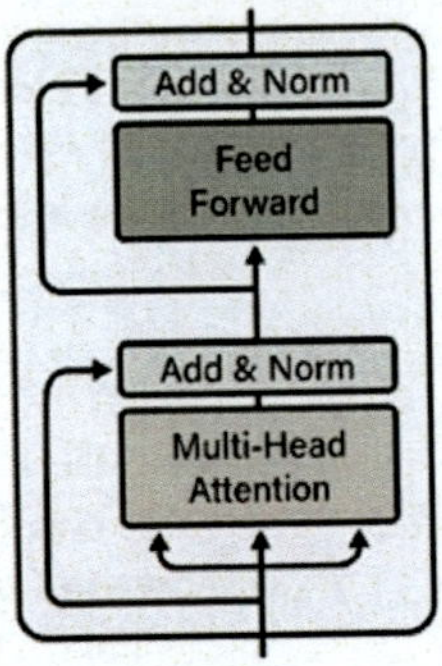

Fig. 2.3 Transformers encoder layer

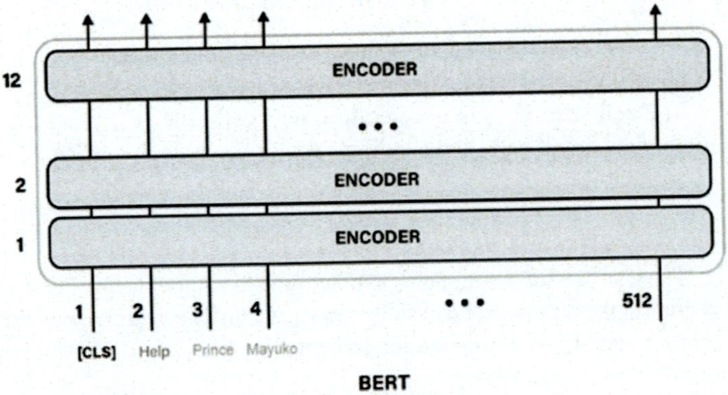

Fig. 2.4 BERT architecture

BERT comprises 12 layers of transformers encoder, each with a hidden size of 768, and the value of □ in the multi-head self-attention layer is 12. The transformer encoder layer consists of two sub-layers in each layer: multi-head attention and position-wise feed-forward network (Fig. 2.4).

Multi-head Attention Multi-head attention is an architecture that simultaneously performs the attention function □ times using different Query, Key, and Value matrices. The goal of multi-head attention is to generate as much as different amounts of attention for each word. As the model processes each word (each position in the input sequence), attention allows it to look at other positions in the input sequence for clues that can help better encode this word.

Position-Wise Feed Forward Network Position-wise feed-forward network is a neural network architecture used to transform the representation of all sequence positions using the same feed-forward network. The feed-forward network architecture consists of two linear transformations with a ReLU (rectified linear unit) activation function between the two linear transformations.

$$\text{FFN} = \max(0, xW1 + b1)W2 + b2 \tag{2.8}$$

With x as the input vector, W_1 as weight matrices from the first layer, W_2 as weight matrices from the second layer, and b as bias.

The BERT model used in this study is IndoBERT-based uncased. IndoBERT-based uncased is the Indonesian version of the BERT model that uses uncased data during pre-training. This model has 12 layers of transformer encoder, 768 hidden sizes, and 12 heads in the attention sub-layer.

2.3 Experiment

In this section, we will describe the process of the experiment. In this study, we will implement continual learning on some domains of Indonesian sentiment analysis using BERT and then compare it to two other models, the fine-tuned embedding with CNN and the fine-tuned embedding with LSTM. The model is trained on personal computer with Intel(R) Core i7, 16GB RAM, an NVIDIA GeForce RTX 3050, and Python 3.7.

2.3.1 Data Sets

There are six data sets used in this study, shown in Table 2.1. *Calon Presiden* contains tweets about the Indonesian Presidential Elections in 2014, while E-commerce contains tweets about e-commerce's existence in Indonesia. Four of the data sets, DANA, Shopback, Grab, and Jenius, have Indonesian reviews about applications from Google Playstore.

Table 2.1 Data sets details

Data sets	Role	Negative sentiment	Positive sentiment	Total sentiments
DANA	Target domain	406	769	1.175
Calon Presiden	Source domain 1	768	1.117	1.885
E-commerce	Source domain 2	422	530	952
Shopback	Source domain 3	979	857	1.836
Grab	Source domain 4	833	755	1.588
Jenius	Source domain 5	943	876	1.819

2.3.2 Preprocessing

There are several changes applied to the text, such as capital letters being changed to lowercase, the website address is removed, the Twitter username deleted, Hashtag removed, punctuation removed, numbers being deleted, the extra spaces being removed, repeating words being separated by removing the dash, letters that are repeated more than two times are deleted into just two times, words with a single letter are removed, and the "*rt*" is deleted. The sentiment labels on the data sets are processed by one-hot encoding. Sentiment on the text has a value of -1 or 1, where -1 represents negative sentiment and 1 represents positive sentiment. Through this preprocessing, a sentiment is mapped into a two-dimensional vector. In the negative sentiment, -1, the mapping result is a vector with the first and second elements 1 and 0, respectively. On the other hand, for the positive sentiment, 1, the mapping result is a vector with the first and second elements being 0 and 1, respectively. In this study, the proportion of training data to testing data is 8:2.

2.3.3 Model Implementation

The first step in the BERT model is to change every word in the sentence input into a numerical vector representation which is then entered into the encoder layer. First, BERT uses the WordPiece model as a tokenizer to tokenize a sentence, and the addition of two special tokens, [CLS] is added at the beginning, and [SEP] is added at the end of the sentence. Padding and truncating are performed to ensure each sentence in the data has the same length of tokens. In this study, the maximum number of tokens is 256. Each document with less than 256 tokens will be padded with a special token [PAD] until the document length reaches 25 tokens, and the sentence with more than 256 tokens will be truncated only up to 256 tokens. The next step is embedding, which functions to map each token to a numeric vector with a particular dimension. Each token has three embeddings, token embedding, segment embedding, and position embedding. The illustration of embedding is shown in Fig. 2.5.

Finally, as shown in Fig. 2.3, the three vectors are added together after obtaining the numerical representation vectors of the token embedding, segment embedding, and positional embedding to obtain the input for the BERT model.

The simulation performs fine-tuning using BERTForSequenceClassification with batch sizes 16, Adam learning method, the learning rate of $2e^{-5}$, and 15 epochs. In this study, the author used early stopping that monitors validation loss with patience =3.

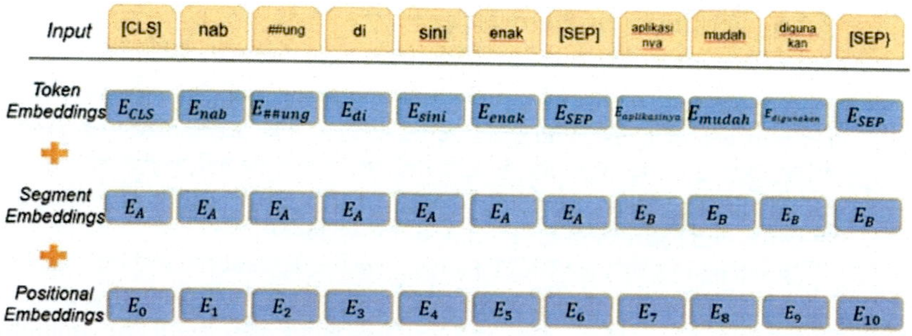

Fig. 2.5 Illustration of input representation for BERT

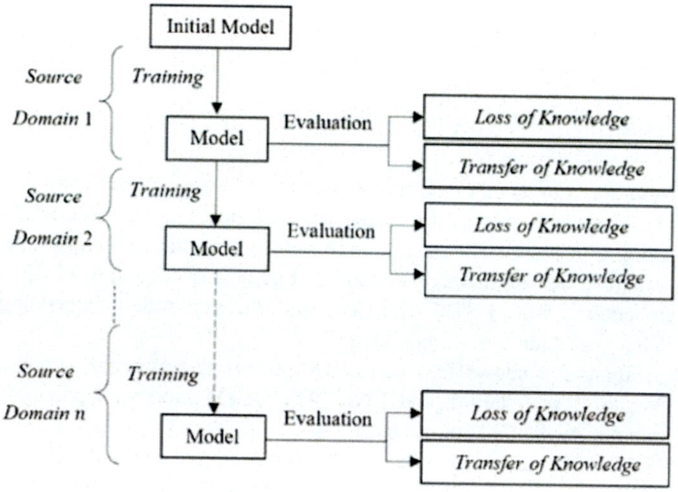

Fig. 2.6 Continual learning implementation process

2.3.4 Continual Learning Implementation

Continual learning is implemented in the model with a flowchart shown in Fig. 2.4 using five data sets. The role of the data sets can be seen in Table 2.1. Based on Fig. 2.4, after the model is built using Source Domain 1, the model continues to learn from Source Domain 2. After learning from Source Domain 2, the model is tested for retain of knowledge (loss of knowledge) by evaluating it to data testing of Source Domain 1. In addition, the model is also tested for the transfer of knowledge by considering it to the Target Domain. The following learning of Source Domains 3, 4, and 5 goes through the same steps (Fig. 2.6).

2.3.5 Transfer of Knowledge

Firstly, we simulate the performance of BERT for continual learning based on the performance of transfer of knowledge, namely, the performance of BERT in the target domain after learning in a series of source domains. Figure 2.7 compares BERT, the fine-tuned embedding with LSTM (LSTM), and the fine-tuned embedding with LSTM (CNN) for transfer of knowledge.

Based on Fig. 2.7, the BERT model increases accuracy to 89.60%. This accuracy increased by 6.6.7% from the initial accuracy of 82.93%, the LSTM model experienced an increase of accuracy to 84.51% or an increase of 19.86% from the initial accuracy of 64.65%, CNN model experienced an increase in accuracy to 85.86%, or an increase of 17.09% from initial accuracy of 68.77%. Based on these results, we can conclude that BERT provides the highest accuracy for the transfer of knowledge.

2.3.6 Retain of Knowledge

Next, we simulate the performance of BERT for continual learning based on retain of knowledge, namely, the performance of BERT in an initial source domain after learning in a series of other source domains. In this simulation, the initial source domain is set to Source Domain 1. Figure 2.8 compares BERT, the fine-tuned embedding with LSTM (LSTM), and the fine-tuned embedding with LSTM (CNN) for retaining of knowledge.

Figure 2.8 shows that the LSTM model retains more knowledge of Source Domain 1 than BERT and CNN. The LSTM model experienced a decrease in accuracy to 83.63% or as much as 2.37% from the initial accuracy of 86.00%, BERT model experienced a reduction in accuracy of up to 80.96% or as much as 8.21% from the initial accuracy of 89.17%, and CNN model experienced a decrease in accuracy to 72.63% or as much as 13.92% from the initial accuracy of 86.55%.

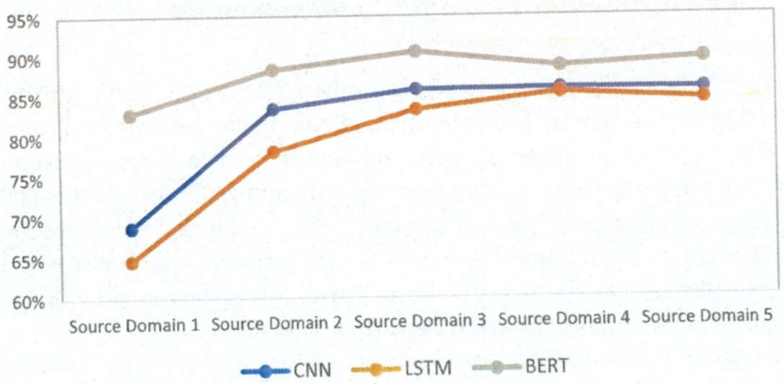

Fig. 2.7 The accuracies of BERT, LSTM, and CNN for transfer of knowledge

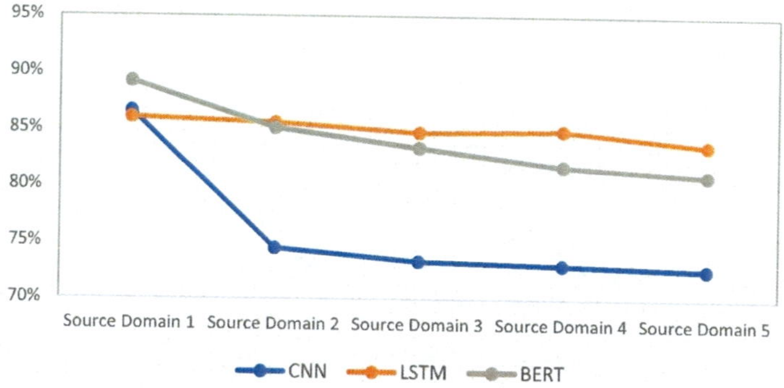

Fig. 2.8 The accuracies of BERT, LSTM, and CNN for retain of knowledge

Table 2.2 The top five highest accuracies of BERT for transfer of knowledge

The sequence of source domain	Accuracy (%)
1-5-4-2-3	91.66
3-1-5-2-4	91.57
5-1-3-2-4	91.49
5-2-3-1-4	91.49
3-5-1-2-4	91.49

Table 2.3 The top five highest accuracies of BERT for retaining knowledge in the source domain of Calon Presiden

The sequence of source domain	Accuracy (%)
1-5-3-2-4	83.29
1-5-2-3-4	81.70
1-2-4-5-3	81.43
1-3-5-4-2	81.42
1-2-3-4-5	80.96

2.3.7 Sequences of Source Domains

Further experiments were conducted on the 120 possible combinations of sequences of the 5 source domains. The experiments aim to see whether or not the order of source domains impacted the accuracy of the BERT model for lifelong learning.

Table 2.2 shows the top five BERT accuracies for transferring knowledge. The highest overall accuracy for transferring knowledge with the BERT model is achieved with the domain sequence of 1-5-4-2-3 and an accuracy of 91.66%. An improvement of 2.06% from an earlier experiment with the sequence of 1-2-3-4-5 that had an accuracy of 89.6%. These simulations show that the order of the source domains affects the performance of BERT for the transfer of knowledge.

We use five scenarios to simulate retain of knowledge, where each source domain becomes the initial source domain. Tables 2.3, 2.4, 2.5, 2.6, and 2.7 show the five highest BERT accuracies for retaining knowledge in each initial source domain. Based on Table 2.3, the highest accuracy of BERT in the source domain Calon

Table 2.4 The top five highest accuracies of BERT for retaining knowledge in the source domain of E-commerce

The sequence of source domain	Accuracy (%)
2-1-4-3-5	90.58
2-3-4-1-5	90.58
2-4-3-1-5	90.58
2-3-1-5-4	90.42
2-1-3-4-5	89.53

Table 2.5 The top five highest accuracies of BERT for retaining knowledge in the source domain of Shopback

The sequence of source domain	Accuracy (%)
3-4-1-2-5	95.65
3-4-1-5-2	95.11
3-1-2-4-5	94.84
3-5-1-4-2	94.84
3-4-5-1-2	94.57

Table 2.6 The top five highest accuracies of BERT for retaining knowledge in the source domain of Grab

The sequence of source domain	Accuracy (%)
4-2-5-1-3	95.28
4-5-1-2-3	93.71
4-1-2-5-3	93.40
4-2-1-5-3	93.08
4-2-5-3-1	92.77

Table 2.7 The top five highest accuracies of BERT for retaining knowledge in the source domain of Jenius

The sequence of source domain	Accuracy (%)
5-3-1-4-2	97.53
5-3-4-1-2	96.43
5-4-1-3-2	96.15
5-1-3-4-2	95.60
5-1-4-3-2	95.60

Presiden was obtained after studying a series of other source domains, with the order of 1-5-3-2-4 being the highest, with an accuracy of 83.29%. In the source domain of E-Commerce, the highest accuracy is 95.65%, provided by the source domain sequence of 3-4-1-2-5 in Table 2.4. For the rest, the source domains of Shopback, Grab, and Jenius, the highest accuracies resulted from the source domain sequences of 3-4-1-2-5, 4-2-5-1-3, and 5-3-1-4-2, respectively.

These simulations also show that the order of the source domains affects the performance of BERT in retaining knowledge. Moreover, there is no correlation between the order of sources domains that give the highest accuracies for both transfer of knowledge and retain of knowledge.

2.4 Conclusion

In this paper, we analyze the performance of the BERT model for lifelong learning in Indonesian sentiment analysis. Then it will be compared with two standard deep learning models: fine-tuned embedding with CNN and fine-tuned embedding with LSTM. Our simulation shows the BERT model gives the best accuracy for the transfer of knowledge. Lifelong learning increases the accuracy by 6.67% from the initial source domain to the last source domain and achieves the final accuracy of 89.60%.

The fine-tuned embedding with CNN model is the second with a final accuracy of 85.86%, followed by the fine-tuned embedding with LSTM with 84.51%. However, the fine-tuned embedding with LSTM model is the best model for retain of knowledge. The fine-tuned embedding with LSTM model's final accuracy is 83.63%, while the BERT model only has a final accuracy of 80.96%. Moreover, our simulation shows that the order of the source domains affects the performance of BERT for both transfer of knowledge and retain of knowledge. There is no correlation between the order of sources domains that give the highest accuracies for both transfer of knowledge and retain of knowledge.

References

1. B. Liu, L. Zhang, A survey of sentiment analysis and opinion mining, in *Mining Text Data*, ed. by C. Aggarwal, C. Zhai, (Springer, Boston, 2012), pp. 413–463
2. B. Liu, Sentiment analysis and opinion mining, in *Synthesis Lectures on Human Language Technologies*, (Morgan & Claypool Publishers, San Rafael, California, 2012)
3. M. Mohri, A. Rostamizadeh, A. Talwalkar, *Foundation of Machine Learning* (MIT Press, Cambridge, 2018)
4. Z. Lipton, M. Li, A. Smola, A. Zhang, Dive into deep learning, https://d2l.ai/. Accessed 01 June 2022
5. I. Goodfellow, Y. Bengio, A. Courville, *Deep Learning* (MIT Press, Cambridge, 2016)
6. Y. Kim, Convolutional neural networks for sentence classification, in *Proceedings of the 2014 Conference on Empirical Methods in Natural Language Processing*, (Association for Computational Linguistics (ACL), Doha, 2014), pp. 1746–1751
7. A. Hassan, A. Mahmood, Deep learning for sentence classification. In: IEEE Long Island Systems, Applications and Technology Conference (LISAT), New York (2017)
8. P.M. Sosa, Twitter sentiment analysis using combined LSTM-CNN models. In: Academia.Edu, pp. 1–9 (2017)
9. T. Gowandi, H. Murfi, S. Nurrohmah, Performance analysis of hybrid architectures of deep learning for Indonesian sentiment analysis, in *Soft Computing in Data Science. SCDS 2021. Communications in Computer and Information Science*, ed. by A. Mohamed, B.W. Yap, J.M. Zain, M.W. Berry, vol. 1489, (Springer, Singapore, 2021), pp. 18–27
10. J. Devlin, M.W. Chang, K. Lee, K. Toutanova, BERT: Pre-training of deep bidirectional transformers for language understanding, in *Proceedings of the 2019 Conference of the North American Chapter of Association for Computational Linguistics: Human Language Technologies*, vol. 1, (Association for Computational Linguistics, Minneapolis, 2019), pp. 4171–4186

11. H. Murfi, T. Gowandi, Syamsuriani, G. Ardaneswari, S. Nurrohmah, BERT-based combination of convolutional and recurrent neural network for Indonesian sentiment analysis. arXiv:2211.05273 [cs.CL] (2022). https://doi.org/10.48550/arXiv.2211.05273
12. G.I. Parisi, R. Kemker, J.L. Part, C. Kanan, S. Wermter, Continual lifelong learning with neural networks: A review. Neural Netw. **113**, 54–71 (2019)
13. C.S. Lee, A.Y. Lee, Clinical applications of continual learning machine learning. Lancet Digit Health **2**(6), e279–e281 (2020). https://doi.org/10.1016/S2589-7500(20)30102-3
14. M. Lenga, H. Schulz, A. Saalbach, Continual learning for domain adaptation in chest x-ray classification. Proc. Third Conf. Medi. Imaging Deep Learn. PMLR **121**, 413–423 (2020)
15. D. Kiyasseh, T. Zhu, D. Clifton, A clinical deep learning framework for continually learning from cardiac signals across diseases, time, modalities, and institutions. Nat. Commun. **12**, 4221 (2021). https://doi.org/10.1038/s41467-021-24483-0
16. T. Ganegedara, *Natural Language Processing with TensorFlow* (Packt Publishing, Mumbai, 2018)

Chapter 3
Multi-scale Dual-Attention-Based U-Net for Breast Cancer Segmentation in Ultrasound Images

Heba Abdel-Nabi , Mostafa Ali , and Arafat Awajan

Abstract Ultrasound imaging is a significant and valuable assistant in reducing breast cancer mortality rate. It has high sensitivity, considered safe and cost-effective. This excellent performance caused the development of a sudden and growing interest in segmenting the breast ultrasound images to distinguish the abnormalities in them. However, this is a very challenging task due to the noisy nature of these images. This paper proposed a Multi-scale Dual-Attention based U-Net framework (MDAU-Net) for automatic breast ultrasound segmentation to address this issue. Our developed framework enhanced the quality of the extracted features by integrating different techniques such as multi-scale input representation, channel attention, spatial attention, blended attention, hybrid pooling and dilated convolutions. The experimental results and the comparison of the proposed model with the state-of-the-art segmentation methods on the well-known widely-used public 'Dataset B' demonstrate the effectiveness and competitiveness of the proposed algorithm.

Keywords Ultrasound · Medical Image Segmentation · Attention

3.1 Introduction

In 2020, (2,261,419) new breast cancer cases were reported worldwide, with (684,996) death cases [1]. According to the global cancer report of the World Health Organization [2], breast cancer is the fourth leading deadly threat worldwide, with a rate of 6.6%. One critical factor that reduces this high mortality rate and increases survivability is the increased awareness of the importance of early detection [3] by

H. Abdel-Nabi (✉) · A. Awajan
Department of Computer Science, Princess Sumaya University for Technology, Amman, Jordan
e-mail: heb20179004@std.psut.edu.jo

M. Ali
Faculty of Computer & Information Technology, Jordan University of Science & Technology, Irbid, Jordan

C.-C. Wang, A. Nallanathan (eds.), *6th International Conference on Signal Processing and Information Communications*, Signals and Communication Technology, https://doi.org/10.1007/978-3-031-43781-6_3

performing periodic examinations. This is necessary to enhance the healing and recovery processes and improve the patients' quality of life. It helps detect cancer early by finding any warning signs or symptoms and thus affects the cure rate by selecting the appropriate treatment. Accordingly, this contributed to decreasing the number of breast cancer deaths to a great extent and achieved a survival rate of 88% after 5 years of diagnosis and 80% after 10 years [4]. It is worth noting that the survival rate depends on the cancer stage discovery and the tumor type and severity.

Medical images are vital and invaluable techniques in breast cancer diagnosis and detection. The most commonly used medical modality in practice is digital mammography (DM) [5] because it does not require a highly skilled operator. Mammography uses a low-dose X-ray to project the inner tissues of the breast. Nevertheless, despite its popularity and wide availability and usage, DM suffers from some limitations that may lead to a relatively high rate of false negatives by its inability to detect breast tumors. For instance, DM is not recommended for young women, pregnant women, or women with dense breasts [6]. Also, because DM involves exposure to X-ray radiation, it cannot be done regularly.

On the other hand, breast ultrasound (BUS) imaging has gained increasing interest from economic and clinical points of view. Among the many advantages of breast ultrasound imaging is the noninvasive, nonradioactive, and painless nature of its imaging procedure. It also provides unique soft tissue information and real-time visualization, which make the taken medical images well tolerated. More importantly, ultrasound has a higher sensitivity than mammography [7], especially for dense breasts commonly found among young women [8]. These dense breast tissues can hide the possible suspicious lesions out of the sight of the radiologist [9] because it has similar X-ray mammography absorption properties as the lesion. Ultrasound detected an additional 1.9–4.2 breast cancers per 1000 women with negative mammograms [10].

Therefore, detecting and identifying the abnormalities, delimiting their boundaries in the ultrasound image, and partitioning it into nonoverlapping distinguished regions, i.e., segmenting the image, are a necessary step in the diagnosis process. However, some limitations exist that bind the segmentation process of breast ultrasound images and make them very challenging, such as the high intra-observer variation rate in US image acquisition due to the different characteristics of the different US devices. Moreover, the interpretation of BUS images is subjective, since it depends on radiologists' technical and clinical experience, especially with the wide variations in tumor appearance in images among different patients, i.e., the shape and size of tumor region in BUS images are irregular. Moreover, BUS images have noisy nature because of the many existing artifacts such as attenuation, speckle noise, intensity inhomogeneity [11], and acoustic shadowing effect [12].

To address these issues, computer-aided diagnosis systems, fuelled by the current artificial intelligence methods such as deep learning, provide supplementary assistance in interpreting and analyzing BUS images. Deep learning mainly tries to overcome the operator dependency issue associated with ultrasound images by replacing the handcrafted discriminative features that depend on radiologists' domain knowledge and skills with self-learned features with different levels of representations.

This paper developed a novel deep learning-based breast ultrasound image segmentation framework called Multi-scale Dual-Attention based U-Net (MDAU-Net) to identify breast tumors efficiently and accurately. A blended attention module is introduced to learn the relative importance of different features with respect to each other by exchanging the edge and smoothness information between them. Furthermore, a public dataset for ultrasound imaging is used in the conducted experiments to have a fair basis for comparison. The results demonstrate that the proposed model improves breast tumor segmentation accuracy. We aim to provide reliable guidance to the medical staff to enhance the level of medical care introduced to the patients.

The remainder of this paper is organized as follows. In Sect. 3.2, some recent BUS segmentation methods are summarized. Our methodology is introduced in Sect. 3.3, including the proposed architecture and the used dataset. Section 3.4 elaborates the experiments setup and the evaluation metrics adopted, in addition to presenting the qualitative and comparative results of the evaluation of the public dataset with our proposed segmentation architecture. Finally, the conclusion and the potential future work are given in Sect. 3.5.

3.2 Related Works

Deep learning-based BUS segmentation methods have witnessed remarkable progress and achieved fruitful results in recent years. However, these studies had some limitations that make room for further enhancements. For example, an improved multi-scaled input attention U-Net model with a novel focal loss function based on the Tversky index was proposed in [13]. The combined focal Tversky loss function aimed to overcome the class imbalance issue by focusing on detecting the low probability classes. The focal loss modified the cross-entropy loss by adding a modulating exponent to prevent the vast number of examples of the majority class from dominating the gradient. Moreover, the Tversky similarity index provided a flexible balance of false positives and false negatives.

The RDAU-NET model was proposed in [14] that replaced the plain neural units in the U-Net with residual units to enhance the edge information. The model used an attention gate module in the skip connections to strengthen the learning of the tumor region by aggregating useful features and suppressing irrelevant background information. Furthermore, a channel-wise attention grid-based encoder-decoder segmentation approach was proposed in [15] that performed grid average pooling on the image that reweights the different feature maps on each grid based on their importance and relevancy. Another modified U-Net based on selective kernels was proposed in [16]. The attention mechanism in the selective kernels accounts for the wide variations in breast tumors since the spatial information of the image at different scales is considered by controlling the size of the receptive fields.

An E-like segmentation approach called CF2-Net was utilized in [17]. It enhanced the backbone U-Net by replacing its skip connections with Fusion Stream Path (FSP) network to integrate coarse to fine information. FSP consisted of four modules for feature integration. The first module was the Atrous Spatial Pyramid Pooling unit that was used to obtain the abundant receptive fields for simultaneously capturing characteristics of target regions with different sizes. The second module was the cascade feature fusion unit that was used to fuse the coarse (low level) and fine (high level) information by contextual transmission strategy. The third module was the edge constraint unit, which was used to remit the blurred edge issue by capturing the edge information. The final module was a one-level U-Net unit, and it was applied to extract the high-level features of the concatenated features from the previous two modules.

3.3 Methodology

3.3.1 The Proposed MDAU-Net Architecture

To precisely identify and delineate tumors in breast ultrasound images, we designed a novel deep learning-based segmentation framework based on a fully convolutional neural network called U-Net architecture [18]. The difference between the proposed model and the backbone U-Net is with the modifications and enhancements introduced to its encoder and decoder blocks to refine the segmentation predictions. The proposed MDAU-Net comprises many building blocks described in detail in the following subsections.

The Overall Proposed MDAU-Net As presented in Fig. 3.1, our U-Net backbone consists of five consecutive levels of encoding and decoding paths with their associated skip connections. In addition to the center level that contains a bottleneck rich-contextual summarization representation, the multiple scales of the input are incorporated to compensate for the lost information in the encoder path because of its successive pooling operations that continuously downsampled the features. They provide the model with the contribution of various semantic contextual information obtained from different spatial levels to achieve better segmentation predictions. These scales are obtained by resizing the input image through the nearest-neighbor interpolation method by scales 2, 4, 8, and 16. Dual attention is used in MDAU-Net in the proposed framework through the use of two attention types to boost the feature discrimination learning. The first type is the Convolutional Block Attention Module (CBAM) [19], used to obtain the features' channel and spatial attention. In contrast, the second type is our novel-designed blended attention block described later.

The Convolutional Building Block This component contains the operation of three sequential convolutions with kernel sizes of (3 × 3), (7 × 7), and (3 × 3) for better adaptation of the feature learning. The block is shown in Fig. 3.2. The activation function is used to provide nonlinearity to the learned features. In contrast,

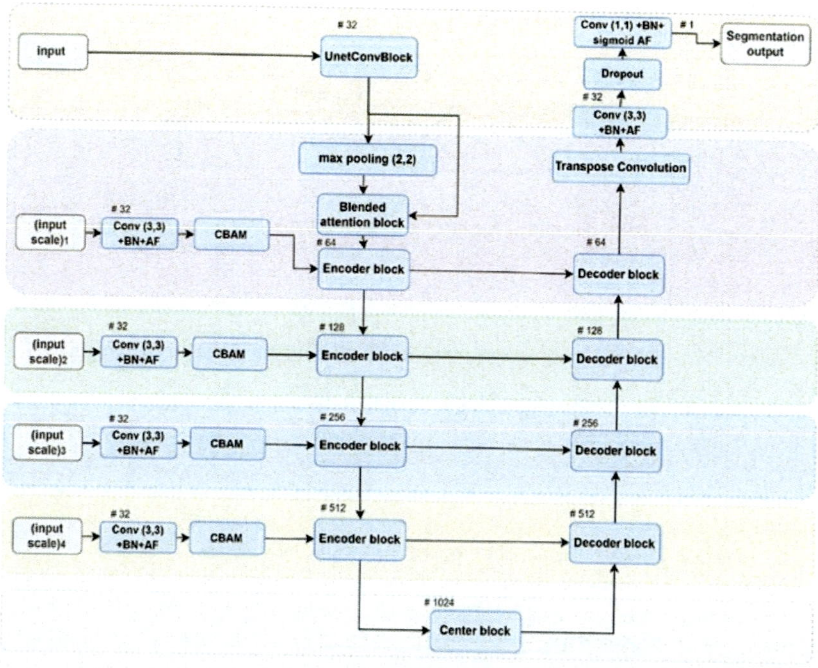

Fig. 3.1 An overview of the proposed MDAU-Net architecture. The numbers above the building blocks indicate the final number of kernels of the outputs of each

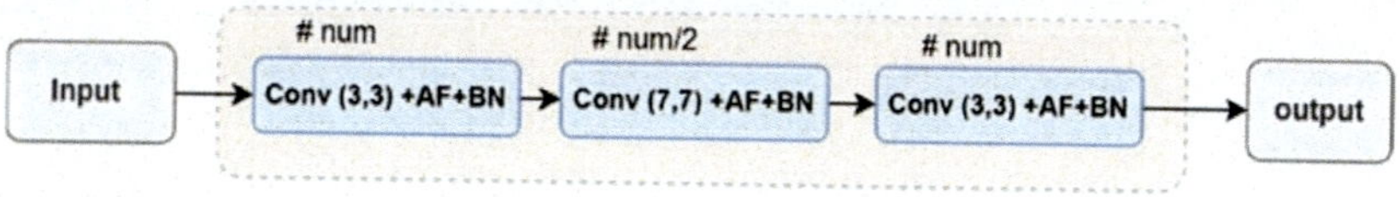

Fig. 3.2 Detailed diagram of the convolutional block (named UnetConvBlock) used in the MDAU-Net, where AF stands for activation function, BN stands for batch normalization, and num is the number of the kernels used in the convolution operation

batch normalization is used to optimize the training better by competing for the required statistics of the data.

The Encoder Block This block takes two inputs: the input scale suitable for this level and the pooling output of the previous level, and produces two outputs – one fed to the next level (layer) and one fed through the skip connection to the same level corresponding decoder, as outlined in Fig. 3.3. The block uses our designed blended attention block to identify the effect caused by the pooling operation on the concatenation of the features extracted after the convolutional block with this level scaled input.

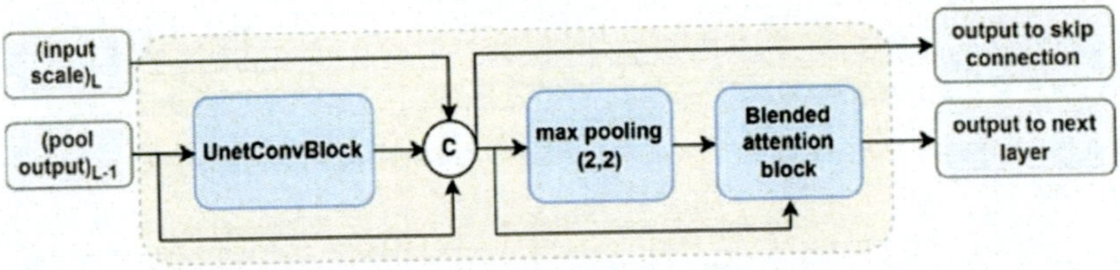

Fig. 3.3 Detailed diagram of encoder block used in the MDAU-Net, where L is the number of the layer (level) of the backbone U-Net and (C) is the concatenation operation

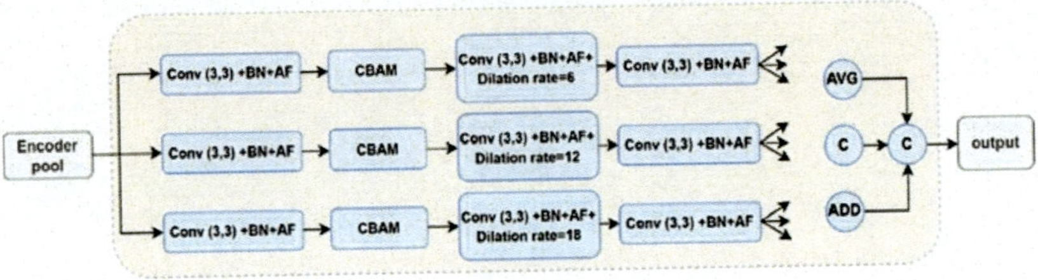

Fig. 3.4 Detailed diagram of the center block in the MDAU-Net

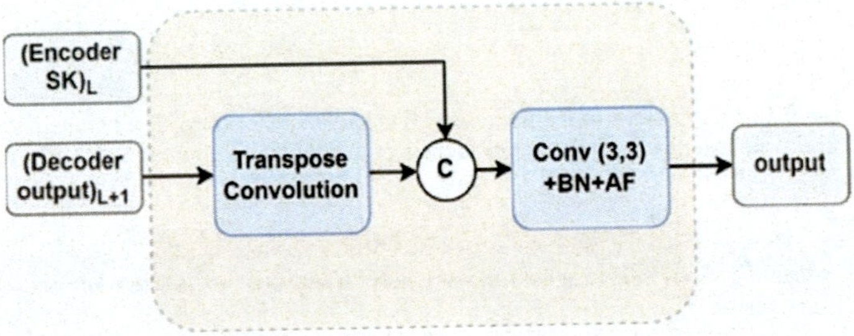

Fig. 3.5 Detailed diagram of decoder block used in the MDAU-Net

The Center Block In order to explore a wide area of the features learned by the encoding path, a dilation convolution is used, instead of the conventional convolution, to prevent the increase of the trainable model capacity. Furthermore, to capture the context at different scales, we use three dilation rates of 6, 12, and 18 in three parallel paths, as shown in Fig. 3.4. Also, we use the CBAM [19] to enhance the feature obtained in each path before the dilated convolution in channel and spatial wise. The concatenation of the three fused features is adopted to learn the various effects of the learned features across the three paths. The triple fusion is done through feature concatenation, element-wise feature averaging, and summation.

The Decoder Block This block takes two inputs: the decoder output of the previous level and the skip connection output of the encoder block, and produces one output that is fed to the next level in the decoding path; see Fig. 3.5. It uses the transpose convolution to upsample the reconstructed features of the previous decoder block.

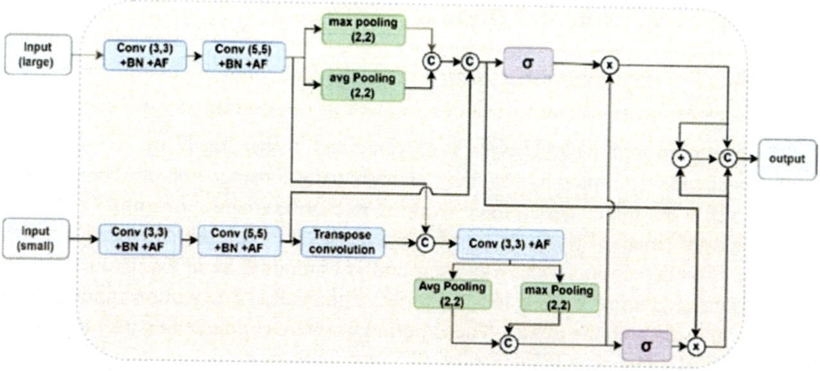

Fig. 3.6 Detailed diagram of blended attention block used in the MDAU-Net, where (x) is the element-wise multiplication and (σ) is the sigmoid activation function

The Blended Attention Block This block takes two inputs of different sizes and produces one output; see Fig. 3.6. The max and average pooling operations are computed on each path but with different positions to extract the edge and spatial information needed to obtain the sharpness and smoothness characteristics, respectively. This hybrid pooling is done to refine the features and thus improve the performance. After that, the sigmoid activation operation is performed on the concatenated vector of the two pooling operations to identify the strength of the feature vectors. Then, an element-wise multiplication is performed between these strength weights and the opposite features of the other path before the sigmoid operation; in other words, a blending between the strength probabilities of each path and the features of the other path is performed. The goal is to exchange the information learned from the two different inputs to highlight the relevant features. Finally, the output is a weighted concatenation of the blended attentions of the two paths in addition to the element-wise summation of their attention coefficient.

3.3.2 *Dataset*

To have a solid, objective, and comparable performance, a public breast ultrasound dataset is used in the evaluation of the proposed model. Our experiments were conducted using the widely used dataset B [20]. The breast ultrasound images in dataset B were collected using Siemens ACUSON Sequoia C512 ultrasound system with 8.5 MHz frequency on 17L5 HD linear array transducer, collected from the Spanish UDIAT Diagnostic Centre of the Parc Tauli Corporation. The BUS images in Dataset B have a mean size of 760 × 570 pixels and contain one or more tumors. Dataset B contains 163 BUS images divided into 53 images with benign tumors and 110 with malignant tumors.

3.4 Experimental Results

3.4.1 Experimental Setup

The proposed MDAU-Net is implemented using Keras on top of TensorFlow backend. The model was trained end to end using stochastic gradient descent (SGD) optimizer with a fixed learning rate of 0.05 and momentum of 0.9. The "He normal" method was used to initialize the layers' weights. The batch size was set to 4. The Dice loss function is used and is computed as in Eq. (3.1). The images in Dataset B were resized to 128 × 128. The "ReLU" activation function introduced nonlinearity to the model. The experiments were conducted on a laptop with a single NVIDIA GeForce RTX 3060 GPU with 6GB memory and Windows 10 operating system. The proposed model was trained for 50 epochs, and the epoch with the lowest validation loss was chosen for the evaluation and prediction at the testing stage. The reported values were taken by averaging the fourfold cross-validation results. To provide a fair comparison base, the comparative methods' results collection also followed fourfold cross-validation.

$$L_{DL} = 1 - \frac{2 * \left(y_{\text{true}} * y_{\text{pred}}\right) + 1}{y_{\text{true}} + y_{\text{pred}} + 1} \tag{3.1}$$

3.4.2 Evaluation Metrics

The quantitative assessment of the validity and effectiveness of our proposed segmentation algorithm of BUS images is evaluated using multiple performance metrics, such as Precision, Recall, and the Dice coefficient. The definitions and the mathematical formulas of these metrics are provided in Table 3.1. True positive (TP) is the correctly segmented number of pixels, i.e., predicted as a tumor pixel that is a true tumor as defined by the radiologist (ground truth tumor). False positive (FP) is the number of pixels that are incorrectly segmented and predicted as a tumor pixel, while it is not a tumor pixel in the manual ground truth. False negative (FN) is the number of pixels incorrectly segmented and predicted as a background (non-tumor) pixel while it is a tumor pixel in the manual ground truth. True Negative (TN) is the correctly segmented number of pixels, i.e., predicted as a background (non-tumor) pixel, while it is also a background pixel in the manual ground truth.

Table 3.1 The used evaluation metrics, their definitions and formulas

Metric	Formula	Definition
Precision	$\text{Precision} = \frac{TP}{TP+FP}$	The ratio of the number of correctly predicted tumor pixels to the total number of predicted tumor pixels in the manual ground truth. The higher the better similarity
Sensitivity ≡ recall	$\text{Recall} = \frac{TP}{TP+FN}$	The ratio of the number of correctly predicted tumor pixels to the total number of true tumor pixels in the manual ground truth. It should be as high as possible in medical image diagnosis. Low value indicates that some tumors are missed and treated as normal tissue
Dice score coefficient (DSC)	$\text{DSC} = \frac{2*TP}{2*TP+FN+FP}$	It measures the overlap or the similarity between the segmented (predicted) tumor region and the true tumor region in the ground truth. The range of the Dice values is [0,1], with 1 indicating the exact similarity and the perfect prediction. The greater the Dice coefficient, the higher the similarity, and thus the better the results

3.4.3 Comparison with the State-of-the-Art BUS Segmentation Methods

In order to validate the effectiveness, competitiveness, and robustness of our proposed segmentation framework, we compared it with state-of-art BUS segmentation methods on the public Dataset B. Table 3.2 summarizes the segmentation results of the proposed MDAU-Net and the eight compared methods using Dataset B. The methods chosen for comparison include basic U-net [18], U-net++ [21], U-net3+ [22], Attention U-net [23], Attention U-net with focal loss [13], RDAU-Net [14], SegNet [24], and SKNet [16]. The comparison was made based on the three metrics discussed earlier: Precision, Recall, and Dice coefficient. Moreover, the loss function used in training each of these models was based on their original papers' recommendations. Our MDAU-Net achieved the best results in the Dice and Precision metrics compared to all the reviewed methods and outperformed them with a good margin. However, our Recall value is slightly lower than the values obtained using U-Net3+ and SegNet due to the class imbalance issue of dataset B and the use of the Dice loss in our training phase compared with the use of cross entropy loss in their training.

3.4.4 Qualitative Results

To evaluate the proposed model efficiency from the qualitative perspective and also to assist the quality of the generated results, a visualization of the segmentation predictions of the proposed model is made in Fig. 3.7, where some examples of the segmentation predictions generated by our proposed model are shown from the

Table 3.2 Quantitative analysis of the segmentation results (mean ± std) of the proposed MDAU-Net and the different methods on Dataset B using average of fourfold cross-validation

	U-Net [18]	U-Net++ [21]	U-Net3+ [22]	Attention U-Net [23]	Attention U-Net/ focal loss [13]	RDAU-Net [14]	SegNet [24]	SKNet [16]	MDAU-Net
Precision	70.2 ± 6.1	68.3 ± 5.7	73.5 ± 6.2	70.4 ± 6.0	73.7 ± 5.1	70.4 ± 4.2	71.7 ± 17	75.3 ± 6.7	**79.2 ± 3.1**
Recall	75.3 ± 2.8	79.6 ± 3.8	80.3 ± 3.9	76.1 ± 4.2	79.2 ± 1.7	73.5 ± 5.2	**80.1 ± 3.9**	79.3 ± 2.5	78.9 ± 3.8
Dice coefficient	68.2 ± 4.2	69.7 ± 5.3	73.9 ± 4.7	69.3 ± 4.1	72.3 ± 3.1	68.2 ± 4.9	72.1 ± 1.5	73.5 ± 4.1	**75.6 ± 4.5**

The best results are highlighted in bold

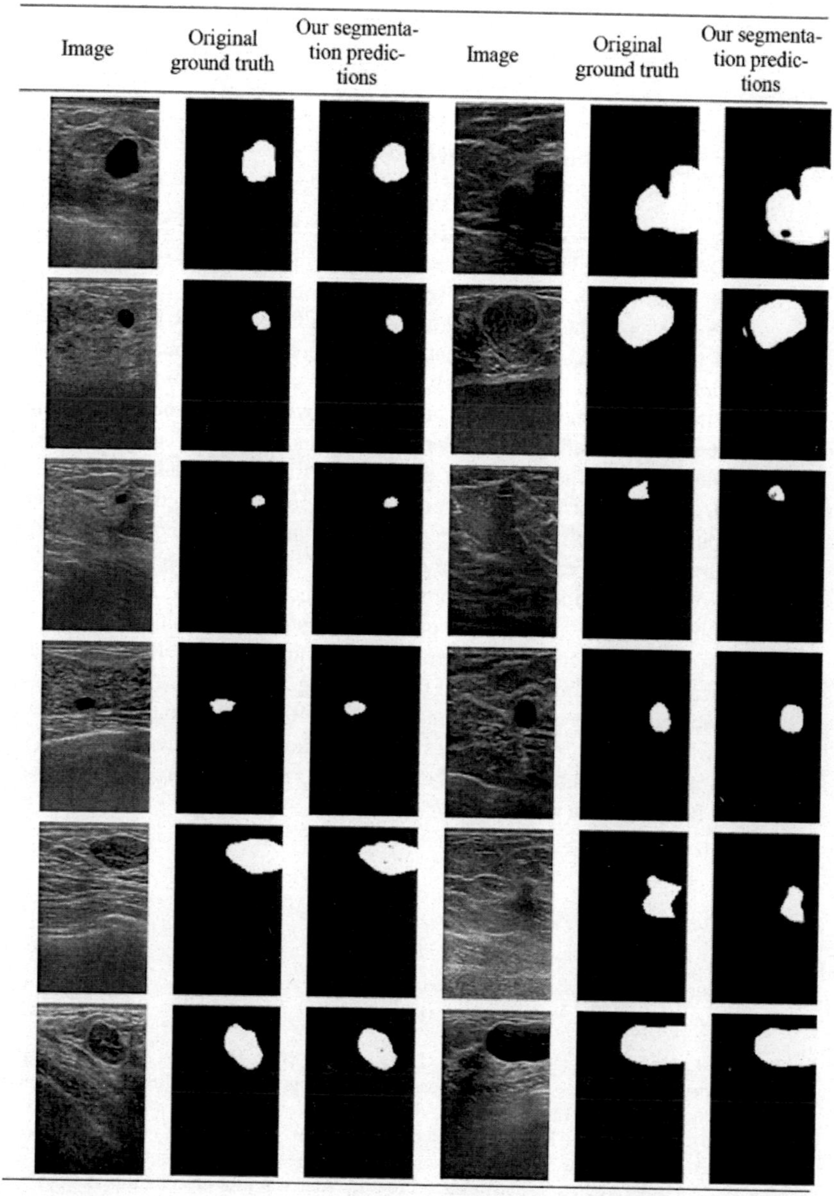

Fig. 3.7 The segmentation results of our proposed segmentation framework for some of the BUS images in the testing set in the Dataset B compared against the ground truth segmentation masks

testing dataset. These BUS images were not fed to the model before for training, indicating the proposed model's good performance and accuracy.

3.5 Conclusion and Future Work

This paper proposed a Multi-scale Dual Attention U-Net-based architecture (MDAU-Net) for breast tumor segmentation in BUS images. Two attention types were integrated into the proposed model to enhance the extracted features from different viewpoints selectively and suppress the irrelevant features. The encoding path was equipped with the input at multiple scales to propagate the spatial information at different levels. Moreover, different kernel sizes with dilated convolutions and three fusion methods were utilized to provide a more robust and representative feature summarization learned by the encoding path at the center of the U-Net. We conducted the experiments on the widely used public dataset B. The proposed MDAU-Net ouperformed the compared state-of-the-art segmentation methods and produced accurate results. Furthermore, qualitative visual segmentation results were reported to better explain the proposed model's performance.

For future work, the performance can be further enhanced by training our proposed segmentation framework with bigger datasets, since the used Dataset B is small in size and does not provide the sufficient data volume required for good utilization of methods based on deep learning techniques. Also, data augmentation techniques can be adopted to increase the robustness of our model. Furthermore, some post-processing techniques can be applied to the generated binary segmentation masks to remove any artifacts and thus reduce our predictions' false-positive and false-negative rates.

References

1. https://gco.iarc.fr/today/. Accessed 5 July 2022
2. https://gco.iarc.fr/today/data/factsheets/cancers/39-All-cancers-fact-sheet.pdf. Accessed 5 July 2022
3. H.D. Cheng, X.J. Shi, R. Min, L.M. Hu, X.P. Cai, H.N. Du, Approaches for automated detection and classification of masses in mammograms. Pattern Recogn. **39**, 646–668 (2006). https://doi.org/10.1016/j.patcog.2005.07.006
4. J. Heymach, L. Krilov, A. Alberg, N. Baxter, S.M. Chang, R.B. Corcoran, W. Dale, A. DeMichele, C.S. Magid Diefenbach, R. Dreicer, Clinical cancer advances 2018: Annual report on progress against cancer from the American Society of Clinical Oncology. J. Clin. Oncol. **36**, 1020–1044 (2018)
5. O. Akin, S.B. Brennan, D.D. Dershaw, M.S. Ginsberg, M.J. Gollub, H. Schöder, D.M. Panicek, H. Hricak, Advances in oncologic imaging. CA Cancer J. Clin. **62**, 364–393 (2012). https://doi.org/10.3322/caac.21156

6. W. Al-Dhabyani, M. Gomaa, H. Khaled, F. Aly, Deep learning approaches for data augmentation and classification of breast masses using ultrasound images. Int. J. Adv. Comput. Sci. Appl. **10**, 1–11 (2019)
7. G. Pons, J. Martí, R. Martí, S. Ganau, J.C. Vilanova, J.A. Noble, Evaluating lesion segmentation on breast sonography as related to lesion type. J. Ultrasound Med. **32**, 1659–1670 (2013). https://doi.org/10.7863/ultra.32.9.1659
8. J.L. Jesneck, J.Y. Lo, J.A. Baker, Breast mass lesions: Computer-aided diagnosis models with mammographic and sonographic descriptors. Radiology **244**, 390–398 (2007). https://doi.org/10.1148/radiol.2442060712
9. M. Yousefi, A. Krzyżak, C.Y. Suen, Mass detection in digital breast tomosynthesis data using convolutional neural networks and multiple instance learning. Comput. Biol. Med. **96**, 283–293 (2018). https://doi.org/10.1016/j.compbiomed.2018.04.004
10. L. Khairunnahar, M.A. Hasib, R.H.B. Rezanur, M.R. Islam, M.K. Hosain, Classification of malignant and benign tissue with logistic regression. Inform Med Unlocked **16**, 100189 (2019). https://doi.org/10.1016/j.imu.2019.100189
11. Q. Huang, Y. Luo, Q. Zhang, Breast ultrasound image segmentation: A survey. Int. J. Comput. Assist. Radiol. Surg. **12**, 493–507 (2017). https://doi.org/10.1007/s11548-016-1513-1
12. A. Madabhushi, P. Yang, M. Rosen, S. Weinstein, Distinguishing lesions from posterior acoustic shadowing in breast ultrasound via non-linear dimensionality reduction, in *2006 International Conference of the IEEE Engineering in Medicine and Biology Society*, (IEEE, 2006), pp. 3070–3073
13. N. Abraham, N.M. Khan, A novel focal Tversky loss function with improved attention U-net for lesion segmentation, in *2019 IEEE 16th International Symposium on Biomedical Imaging (ISBI 2019)*, (IEEE, 2019), pp. 683–687
14. Z. Zhuang, N. Li, A.N. Joseph Raj, V.G.V. Mahesh, S. Qiu, An RDAU-NET model for lesion segmentation in breast ultrasound images. PLoS One **14**, e0221535 (2019). https://doi.org/10.1371/journal.pone.0221535
15. H. Lee, J. Park, J.Y. Hwang, Channel attention module with multi-scale grid average pooling for breast cancer segmentation in an ultrasound image. IEEE Trans. Ultrason. Ferroelectr. Freq. Control **1–1** (2020). https://doi.org/10.1109/TUFFC.2020.2972573
16. M. Byra, P. Jarosik, A. Szubert, M. Galperin, H. Ojeda-Fournier, L. Olson, M. O'Boyle, C. Comstock, M. Andre, Breast mass segmentation in ultrasound with selective kernel U-net convolutional neural network. Biomed Signal Process Control **61**, 102027 (2020). https://doi.org/10.1016/j.bspc.2020.102027
17. K. Wang, S. Liang, S. Zhong, Q. Feng, Z. Ning, Y. Zhang, Breast ultrasound image segmentation: A coarse-to-fine fusion convolutional neural network. Med. Phys. **48**, 4262–4278 (2021). https://doi.org/10.1002/mp.15006
18. O. Ronneberger, P. Fischer, T. Brox, U-net: Convolutional networks for biomedical image segmentation, in *Medical Image Computing and Computer-Assisted Intervention (MICCAI)*, LNCS, vol. 9351, (Springer, 2015), pp. 234–241
19. S. Woo, J. Park, J.-Y. Lee, I.S. Kweon, CBAM: Convolutional block attention module. In: Proceedings of the European Conference on Computer Vision (ECCV) (2018)
20. M.H. Yap, G. Pons, J. Marti, S. Ganau, M. Sentis, R. Zwiggelaar, A.K. Davison, R. Marti, Automated breast ultrasound lesions detection using convolutional neural networks. IEEE J. Biomed. Health Inform. **22**, 1218–1226 (2018). https://doi.org/10.1109/JBHI.2017.2731873
21. Z. Zhou, M.M.R. Siddiquee, N. Tajbakhsh, J. Liang, UNet++: Redesigning skip connections to exploit multiscale features in image segmentation. IEEE Trans. Med. Imaging **39**, 1856–1867 (2020). https://doi.org/10.1109/TMI.2019.2959609
22. H. Huang, L. Lin, R. Tong, H. Hu, Q. Zhang, Y. Iwamoto, X. Han, Y.-W. Chen, J. Wu, UNet 3+: A full-scale connected UNet for medical image segmentation, in *ICASSP 2020 – 2020 IEEE International Conference on Acoustics, Speech and Signal Processing (ICASSP)*, (IEEE, 2020), pp. 1055–1059

23. O. Oktay, J. Schlemper, L.L. Folgoc, M. Lee, M. Heinrich, K. Misawa, K. Mori, S. McDonagh, N.T. Hammerla, B. Kainz, B. Glocker, D. Rueckert, Attention U-Net: Learning where to look for the pancreas (2018)
24. V. Badrinarayanan, A. Kendall, R. Cipolla, SegNet: A deep convolutional encoder-decoder architecture for image segmentation. IEEE Trans. Pattern Anal. Mach. Intell. **39**, 2481–2495 (2017). https://doi.org/10.1109/TPAMI.2016.2644615

Chapter 4
Introduce the CH Role Rotation Mechanism in the Multilayered Deterministic WSN Clustering to Achieve Long-Term Load Balancing

Othmane Dergaoui, Youssef Baddi, and Abderrahim Hasbi

Abstract WSN clustering is a topology management mode which consists of grouping nodes into clusters according to a similarity criterion which is generally geographic proximity. This technique seeks to resolve several problems relating to the operation of WSNs, in particular high energy consumption and unbalanced load distribution. In this context, the multilayered deterministic WSN clustering was set up, and it aims at load balancing while reducing overall energy consumption. Admittedly, the technique envisaged allows to achieve an interesting level of load balancing between CHs over a single communication round, but the unbalanced distribution of loads between all the nodes of the network over several communication rounds is a problem which persists given that CHs consume more energy than member nodes. In this paper, we propose a CH role rotation mechanism in order to carry out long-term load balancing between all the nodes of the WSN. The simulations done have shown that the CH role rotation mechanism can considerably increase the lifetime of the clustered WSN without adversely affecting the quality of the service provided to the end user.

Keywords WSN · WSN clustering · Clustering · Multilayered clustering · CH role rotation · Load balancing

4.1 Introduction

The provision of services through WSNs comes up against several technical problems such as high energy consumption, the hot spot problem linked to the inequitable distribution of loads, network congestion, packet loss, etc.

O. Dergaoui (✉) · A. Hasbi
Mohammed V University, Rabat, Morocco

Y. Baddi
STIC Lab, Chouaib Doukkali University, El Jadida, Morocco

C.-C. Wang, A. Nallanathan (eds.), *6th International Conference on Signal Processing and Information Communications*, Signals and Communication Technology, https://doi.org/10.1007/978-3-031-43781-6_4

[1, 2]. Consequently, clustering has become more and more used as a topology management mode, since it allows to reduce the effect of the abovementioned problems on the quality of the services offered to the end user.

Clustering techniques have evolved over the years, thanks to the opening up of WSN management science to other areas such as data analysis, Big Data, Data science, etc. [2]. Currently, clustering paradigms are based on powerful algorithms from the field of statistical data analysis such as k-medoids. Currently, hundreds of techniques exploit the strength, scalability, and quality of k-medoids in order to perform WSN clustering. Among the recent techniques of clustering based on k-medoids, the multilayered deterministic WSN clustering [3].

This clustering mode made it possible to reach an unprecedented level of load balancing between CHs over a single round of communication while reducing overall energy consumption. However, the clustering proposed in [3] does not guarantee long-term load balancing (over several rounds of communication) between all the nodes of the WSN, whether they are CHs or member nodes. Indeed, in all clustering techniques, the CHs consume much more energy than member nodes on a single round of communication. This gap in energy consumption becomes larger and larger after several rounds of communication, which leads to the hot spot problem which is synonymous with poor load balancing. To remedy this trouble, the CH role rotation mechanism is introduced in clustering techniques. It allows to distribute the energy-intensive tasks accomplished by the CHs on all the nodes of the WSN, in turn after a fixed or variable frequency. Thanks to this mechanism, all the WSN nodes perform, at the end of the WSN operation after several rounds of communications, the tasks of CH and member node in an equitable manner, which leads to a quasi-perfect load balancing. In this context, the CH role rotation paradigm presented in this paper was considered to complement the work done in [3] in order to achieve an almost perfect clustering in terms of load balancing. The paper is organized as follows: Sect. 4.2 describes some general concepts related to the CH role rotation in WSN clustering, Sect. 4.3 contains the state of art of the similar works carried out in recent years, and Sect. 4.4 presents in detail the proposed CH role rotation mechanism. In conclusion, Sect. 4.5 summarizes the contribution of the work carried out in this paper.

4.2 Background and Terminology

4.2.1 WSN Clustering

WSN clustering is a grouping of sensor nodes in groups called "clusters." In each cluster, a node called "cluster head" (CH) is responsible for collecting the packets from other nodes called "member nodes" and relaying them to the next hop. The goal of WSN clustering is to reduce overall energy consumption by minimizing communication distances.

4.2.2 CH Role Rotation

CH role rotation (CRR) is a mechanism used in clustered WSNs. It distributes the CH role which consists of collecting, merging, aggregating, and relaying packets to all sensor nodes. The importance of the CRR is manifested in the great difference in the amount of energy dissipated in the CHs relative to the member nodes. Then, a clustered WSN where a special sensor node always assumes the CH role presents the risk of the hot spot problem [1]. Therefore, the CRR is an obligation to ensure long-term load balancing [1].

CRR can be carried out with or without re-clustering. The choice of which approach to adopt depends on the nature of the services offered by the clustered WSN. However, the approach without re-clustering is the most used, since it is easy, less expensive in time and resources, and does not disrupt the functioning of the WSN. On the other hand, re-clustering takes the WSN out of service at a frequency that can adversely affect QoS. Certainly, the CRR without re-clustering is not ideal in terms of energy consumption since some nodes which are not at the center of clusters will be elected as CHs, but the overall energy consumed will be balanced between all nodes after several rounds of communications. The option without re-clustering presents the best compromise between energy consumption, long-term load balancing, and the availability of the services provided to the end users. This criterion is the supreme purpose of all work done at the level of WSNs.

4.3 State of Art

CRR is an idea that has been in place since the emergence of WSN clustering techniques used in IoT infrastructures realizing sustainable applications and services. This principle is based on the fact that maintaining the CH role for a well-defined group of sensor nodes creates the hot spot problem, especially in clusters close to the sink node [6]. Now, the overwhelming majority of well-known clustering techniques adopt the principle of CRR in order to achieve long-term load balancing.

Conventional clustering techniques such as LEACH, PEACH, and EEUC perform the CRR based on the time criterion (periodicity in number of communication rounds). With the evolution of clustering paradigms, the CRR has also known the appearance of the event criterion where the handover of the CH role is triggered by some remarkable events such as the state of the nodes, the congestion of the bandwidth, etc. [1]. In this context, the ECDR technique [5] was designed to perform a CRR triggered when the battery level of a CH node goes below a well-defined threshold. SEP [7], for its part, is inspired by the probabilistic model of LEACH by adding residual energy as a determining factor in the election of CH nodes.

There are some other techniques that realize the CRR after each round of communication such as Efficient Cluster Radius and Transmission Ranges in

Corona-based Wireless Sensor Networks [8] where the network is organized as a multilevel crown. The intercluster communication is done through multi-hop exchanges between CHs belonging to adjacent levels until the packets reach the sink node.

For techniques based on data analysis techniques, the MK-means paradigm [9] performs a partial CRR where three nodes are chosen in each cluster in order to ensure the CH role in turn. This technique is used especially in heterogeneous WSNs, since the distribution of the CH role is not done equitably, which puts the WSN at the risk of hot spot problem [10] or network congestion in the case where the nodes are homogeneous.

4.4 Implementation of the CH Role Rotation

4.4.1 Calculating the Number of Rounds to Rotate CHs

The CRR approach chosen in this contribution is without re-clustering. All the CHs of the different clusters are modified simultaneously after a number of communication rounds equal to *roundRotate*. *roundRotate* is therefore the periodicity of the CRR in the whole clustered WSN. The total duration of network operation is equal to *numberRounds* communication rounds. The two values of *roundRotate* and *numberRounds* are determined empirically. For our case, we fixed the two aforementioned quantities, respectively, at 50 and 8000 rounds of communication.

4.4.2 The CH Role Rotation Algorithm

The WSN is first of all clustered using the multilayered deterministic WSN clustering algorithm presented in [3]. The latter returns *chs* and *clusteredDataset* which are, respectively, a matrix where each row contains the CHs of a layer, and a multidimensional array containing the 12 clustered layers, each layer contains a number of clusters equal to *clusterNumber*, and each cluster is an array whose first element is the CH. Then, the algorithm (1) operates on *chs* and *clusteredDataset* in order to perform the CRR and evaluate the lifetime of the clustered WSN by calculating the number of alive nodes after each round of communication launched.

The *CRR* function used in algorithm (1) is presented in algorithm (2). It allows to modify the CHs in a circular manner in all the clusters of the clustered WSN.

4.4.3 Routing

The routing is carried out via the two algorithms presented in [3]. These algorithms allow to trace a path of three hops from a CH belonging to layer 3 to a node of layer 0 passing through layers 2 and 1.

4.4.4 Simulation

Algorithm 1 The WSN operating algorithm

```
Input : clusteredDataset, chs, numberRounds, roundRotate,
clusterNumber, L0 (the list of nodes in layer 0)
Output : nodesAlive
  for round is 1 to numberRounds do
    if mod(round, roundRotate) = 0 then
      [clusteredDataset, chs] ← CRR(clusteredDataset, chs) chs ←
routing(chs)
        for i is 5 to 12 do
        for j is 1 to clusterNumber do
          clusteredDataset[i][j][1] ← chs[i][j] chs = routingL01(chs,
L0)
        for i is 1 to 4 do
        for j is 1 to clusterNumber do
          clusteredDataset[i][j][1] ← chs[i][j] numberOfRotations ←
numberOfRotations + 1
     [clusteredDataset, L0] ←
     communicationRound(clusteredDataset, L0) nodesAlive[round] ←
  computeNodesAlive(clusteredDataset, L0)
```

Algorithm 2 The CRR function

```
function CRR(clusteredDataset, chs) clusterNumber ← size
(clusteredDataset[1]) newClusteredDataset ← [][][]
  newCHs ← [][]
    for i is 1 to 12 do
    for j is 1 to clusterNumber do
      sz ← size(clusteredDataset[i][j]) M ← []
      M[1] ← clusteredDataset[i][j][sz] newCHs[i][j] ← M[1]
        for k is 1 to sz-1 do
          M[k+1] ← clusteredDataset[i][j][k] M[k+1].nextHop ← M[1]
      newClusteredDataset[i][j] ← M
  return [newClusteredDataset, newCHs]
```

The choice of the evaluation approach is not made at random; it is based on the nature of the need to which the clustered WSN responds. Indeed, the detection applications require the operation of all the sensor nodes, and the depletion of a node puts all the WSN almost out of service. Consequently, clustering in WSNs intended for detection aims to slow down the death of the first node, which poses load

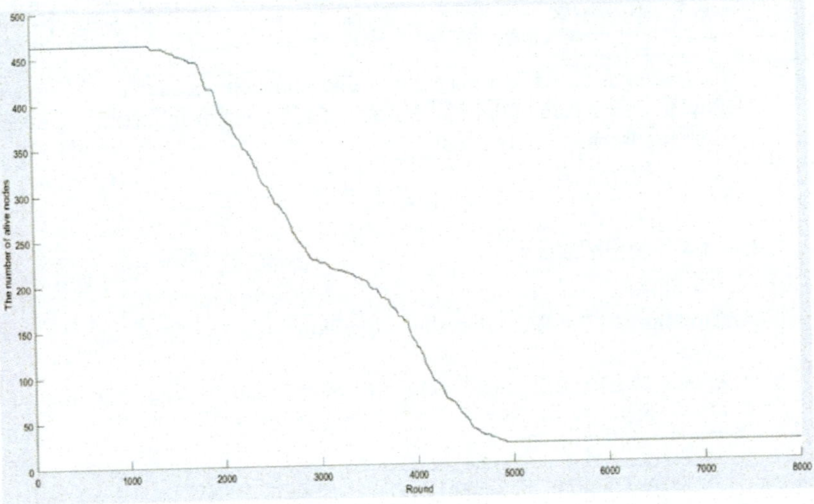

Fig. 4.1 Number of alive nodes per round for the multilayered deterministic WSN clustering with CRR

balancing as the fundamental criterion to be satisfied. On the other hand, monitoring applications require the operation of a percentage of nodes. Therefore, reducing energy consumption is the primordial equation to be solved in the case of monitoring even to the detriment of the overall load balancing of the WSN.

In our case, we will set up the CRR mechanism without re-clustering in all the WSN clusters. Then, we will calculate the two values of PNA and FND before establishing a comparative assessment with some other clustering approaches based on the two aforementioned criteria.

Methods The simulations are carried out in Matlab on a homogeneous WSN composed of 464 nodes deployed on a surface of 100 m * 100 m. These nodes were clustered according to the multilayered deterministic WSN clustering paradigm presented in [3]. This clustering is tested on *numberRounds* rounds of communication with a CRR periodicity equal to *roundRotate*. The simulations compare the evolution of the number of alive nodes per communication round for the multilayered deterministic WSN clustering and some other similar clustering techniques such as EDCR, LEACH, SEP, and ANTCLUST presented in [4].

Results Figure 4.1 shows that the operation of the communication rounds in our clustered WSN goes through three phases, namely:

- The operation phase of the entire network between the first and the round 1231 which represents the FND for the proposed clustering.
- The almost linear depletion phase between the rounds 1232 and 4295 where the network loses 1 node every 9 rounds of communication.

- The phase of almost permanent operation starting from the round 4296. In this phase, the network guarantees the operation of all the services having the capacity to operate with only 22 alive nodes. This phase gives a PNA value equal to 5% for the proposed clustering.

Simulations on other approaches such as EDCR, LEACH, SEP, and ANTCLUST performed in [4] gave the following results:

- EDCR has an FND value close to 1600, and all nodes burn out after 2200 rounds.
- SEP has an FND value close to 1300. Subsequently, the network loses more than 90% of the nodes between the rounds 1300 and 4500. The PNA value for SEP is almost equal to 10%.
- LEACH and ANTCLUST have almost the same evolution curve with almost the same value of FND close to 1000. The only difference between these two techniques is the round from which the WSN is completely exhausted. This value is equal to 2100 for LEACH and almost 3450 for ANTCLUST.

Discussion The EDCR technique achieves, thanks to its high value of FND, an almost perfect level of load balancing. In fact, the near-simultaneous exhaustion of all the nodes proves that they consume almost the same amount of energy after each round of communication. On the other hand, the depletion of the entire WSN from round 2200 makes EDCR an inappropriate technique especially for applications and services which can function even with a reduced number of alive nodes. Therefore, EDCR is the most compatible clustering mode for applications that require the collective operation of all network nodes. However, for a WSN used in an IoT infrastructure that is scalable and open to several types of services, EDCR risks being an unsuitable discriminating choice for uses that rely on minimal operation for a long duration.

LEACH and ANTCLUST have a PNA equal to 0% and a reduced value of FND. These two techniques are then penalizing for any service or application used for a medium or long period. Then, the two aforementioned techniques can be the subject of a performance analysis for the WSN clustering techniques used in on-demand or short-use applications.

SEP achieves similar performances to LEACH and ANTCLUST for applications using all the nodes of the network since the three aforementioned clustering techniques have almost the same value of FND. Nevertheless, the value of PNA close to 10% makes SEP more useful than LEACH and ANTCLUST especially for sustainable applications that can operate with a reduced set of sensor nodes.

The technique proposed in this paper achieves the best compromise between FND and PNA values. Admittedly, there are more efficient techniques for some particular cases, notably EDCR for applications requiring the collective operation of all the sensor nodes in a medium or short time, but the multilayered deterministic WSN clustering equipped with the CRR mechanism is the most adaptive as it gives interesting performances for the overwhelming majority of IoT applications, thanks to its FND and PNA values which are very close to the average of all the WSN clustering techniques used at this time. Therefore, in an open and scalable IoT infrastructure, the multilayered deterministic WSN clustering with CRR is the most judicious option, since flexibility and adaptability are the most required criteria in the choice of the network topology management modes.

4.5 Conclusion

Clustering is the most widespread topology management mode in the world of WSNs given its efficiency, performance, and ability to solve the challenges related to the use of WSNs such as high energy consumption, load balancing, network congestion, QoS, etc. However, pure clustering offers a partial solution to the problem of load balancing, since it allows to balance the quantities of energy consumed by the sensor nodes over a single round of communication without offering a durable load balancing. In this context, the CH role rotation has been designed as an interesting extension performing long-term load balancing, which aligns with the requirements of a large family of IoT applications and services operating for long durations. The evaluation of long-term clustering paradigms is done via two values calculated in number of communication rounds, namely, the FND and the PNA. The quantitative evaluation according to the two aforementioned criteria makes it possible to match each clustering technique to the appropriate family of IoT applications. In this context, the quantitative evaluation of the multi-layered deterministic WSN clustering with CRR shows that this technique has reached, thanks to the almost perfect and exclusive balance between the high value of PNA and FND, an unprecedented level of adaptability and flexibility.

References

1. A. Shahraki, A. Taherkordi, Ø. Haugen, F. Eliassen, Clustering objectives in wireless sensor networks: A survey and research direction analysis. Comput. Netw. **180**, 107376 (2020). https://doi.org/10.1016/j.comnet.2020.107376
2. Y. Gong, J. Wang, G. Lai, Energy-efficient query-driven clustering protocol for WSNs on 5G infrastructure. Energy Rep. **8**, 11446–11455 (2022). https://doi.org/10.1016/j.egyr.2022.08.279
3. O. Dergaoui, Y. Baddi, A. Hasbi, Energy-saved and load-balanced wireless sensor network clustering in a multi-layered wireless networks structures. In: 2022 5th Conference on Cloud and Internet of Things (CIoT), pp. 122–128 (2022). https://doi.org/10.1109/CIoT53061.2022.9766675
4. S. Gamwarige, C. Kulasekere, An algorithm for energy driven cluster head rotation in a distributed wireless sensor network. Proceedings of the International Conference on Information and Automation, December 15–18, Colombo, Sri Lanka (2005)
5. M. Lewandowski, B. Płaczek, An event-aware cluster-head rotation algorithm for extending lifetime of wireless sensor network with smart nodes. Sensors **19**(4060) (2019). https://doi.org/10.3390/s19194060. Author, F.: Contribution title. In: 9th International Proceedings on Proceedings, pp. 1–2. Publisher, Location (2010)

6. S.K. Singh, P. Kumar, J.P. Singh, An energy efficient protocol to mitigate hot spot problem using unequal clustering in WSN. Wirel. Pers. Commun. **101**, 799–827 (2018). https://doi.org/10.1007/s11277-018-5716-3
7. N.R. Roy, P. Chandra, Threshold sensitive clustering in SEP. Sustain. Comput. Inform. Syst. **25**, 100367 (2020). https://doi.org/10.1016/j.suscom.2019.100367
8. W.K. Lai, C.-S. Fan, C.-S. Shieh, Efficient cluster radius and transmission ranges in corona-based wireless sensor networks. KSII Trans. Internet Inf. Syst. **8**, 1237–1255 (2014)
9. S. Periyasamy, S. Khara, S. Thangavelu, Balanced cluster head selection based on modified k-means in a distributed wireless sensor network. Int. J. Distrib. Sensor Netw. **12**, 5040475 (2016). https://doi.org/10.1155/2016/5040475
10. R. Elkamel, A. Messouadi, A. Cherif, Extending the lifetime of wireless sensor networks through mitigating the hot spot problem. J. Parallel Distrib. Comput **133**, 159–169 (2019). https://doi.org/10.1016/j.jpdc.2019.06.007

Chapter 5
IoT Feature Assessment for Smart Cities via Intuitionistic Fuzzy Selected Element Reduction Approach (IF-SERA)

Esra Çakır and Emre Demircioğlu

Abstract IoT technologies implemented into congested cities surely improve society's quality of life. However, it is critical to accurately define the criteria for the adaptations of cities in this process and how to make an evaluation. While smart cities are graded based on assessment criteria such as energy savings, data quality and integrity, cloud computing, and management issues, it is also vital to consider which features will be effective and how much they will rate. This study evaluates IoT features of being smart cities by integrating Intuitionistic Fuzzy Selected Element Reduction Approach (IF-SERA), which is newly introduced to the literature for weighting criteria in decision-making problems in an intuitionistic fuzzy environment. This method is based on the impact of a chosen criterion on the findings, and then the weight is determined by eliminating it from the evaluation. Hence, while making an assessment, the criteria weights are not subjective, and they take evaluation-specific values. Within the scope of smart cities, IoT criteria are ranked according to intuitionistic fuzzy decisions of the experts. With the case of IoT-based smart cities, the application area of the novel fuzzy weighting approach is investigated.

Keywords Fuzzy multi-criteria decision-making · Intuitionistic fuzzy set · IoT-based smart cities · Selected Element Reduction Approach

5.1 Introduction

In congested cities, more communication of each unit improves life quality. The Internet of Things (IoT) plays an essential role in communication, from controlling small domestic appliances to communicating with massive metropolitan structures [1]. The capacity to remotely operate things and communicate real-time data hastened the sequence of events. Thus, the use of smart solutions in cities ensures that

E. Çakır (✉) · E. Demircioğlu
Department of Industrial Engineering, Galatasaray University, İstanbul, Turkey
e-mail: ecakir@gsu.edu.tr

C.-C. Wang, A. Nallanathan (eds.), *6th International Conference on Signal Processing and Information Communications*, Signals and Communication Technology, https://doi.org/10.1007/978-3-031-43781-6_5

society can communicate and act in a variety of sectors such as transportation, air quality, and waste management [2]. Smart cities are evolving on a daily basis in order to provide a sustainable environment and a healthy lifestyle. In addition to the emphasis cities place on smart solutions, improving their quality of life influences their evaluations [3].

In this study, a new fuzzy approach SERA [4] is applied in the evaluation of smart cities in line with the linguistic expressions of the experts. Extending Zadeh's [5] concept of fuzzy numbers as intuitionistic fuzzy numbers (IF), Atanassov [6] proposed a new fuzzy number for the use of IF numbers. Using the terms of membership and nonmembership in the same expression, he inspired many decision-making approaches [7–10]. There are also numerous IoT and smart city studies that use fuzzy numbers and are considered as decision-making problems in the literature. Seker [11] evaluated an IoT-based sustainable smart waste management system with interval-valued q-rung orthopair fuzzy MCDM methodology. He applied a modified Entropy approach for the local municipality in Istanbul. Ozkaya and Erdin [12] performed a hybrid ANP and TOPSIS methodology on smart cities. According to their assessment, among 44 cities, Tokyo, London, and Boston were the cities with low quality-of-life scores, and they were recommended to invest more in smart city adaptation. Sharma et al. [13] explored the barriers of waste management in smart cities for India to IoT adoption. According to their results, the MCDM-based model performed approximately 50% better than the non-MCDM models. Lin et al. [14] proposed a new integrated probabilistic linguistic MCDM model based on the TODIM method to rank IoT platforms, based on the probabilistic linguistic best-worst (PLBW) method, two-tuple distance measure, and two-level possibility degree to evaluate IoT platforms. Büyüközkan and Uztürk [15] are stressed on smart last-mile delivery solutions. They performed an in-depth evaluation of last-mile delivery from a smart city perspective and proposed a 2-Tuple integrated DEMATEL-VIKOR methodology for the Istanbul case. Rajab and Cinkelr [16] investigated the primary difficulties and shortcomings of using IoT technologies based on smart city principles. Based on these case studies, it is also an important step to determine the weights of the criteria to be used when evaluating smart cities. Therefore, in this study, the recently proposed IF-SERA [17], which produces its own criterion weight instead of using subjective evaluations, is preferred. In the literature, some applications performed using the SERA method are service quality for digital suppliers [4] and personnel selection case [17] in fuzzy environments.

This study contributes to the literature by recommending the use of IF-SERA for the case of IoT features evaluation of smart cities. Intuitionistic fuzzy sets are in a format suitable for combining viewpoints in group decision-making. In addition to the applications of SERA, which is a fuzzy multi-criteria decision-making technique proposed by the authors of this paper, a IF integration has been carried out. IoT features are evaluated using the decision-makers' fuzzy linguistic IFS decisions, and then SERA procedure is applied. The proposed approach is used in a ranking of several IoT features of being smart cities to assess their adaptation to be smart.

The paper is designed as follows. Section 5.2 gives the preliminaries on intuitionistic fuzzy sets. Section 5.3 presents the proposed IF-SERA methodology in detail. Section 5.4 performs the proposed methodology on IoT features assessment for smart cities. Finally, conclusions and future directions are discussed in Section 5.5.

5.2 Preliminaries

After Zadeh's presentation of fuzzy sets in 1965 [5], Atanassov [6] expanded the fuzzy idea to provide the intuitionistic fuzzy set framework, which takes into consideration the degree of membership and nonmembership. The total of membership and nonmembership degrees in this type of fuzzy expression cannot be more than one. The following are definitions of intuitionistic fuzzy sets:

Definition 5.1 [18] Let $X = \{x_1, x_2, \ldots, x_n\}$ be a universe of discourse, an intuitionistic fuzzy set (IFS) A in X is given by

$$A = \{ < x, u_A(x), v_A(x) > x \in X\} \tag{5.1}$$

where $u_A : X \to [0, 1]$ and $v_A : X \to [0, 1]$ with the conditions $0 \leq u_A(x) + v_A(x) \leq 1, \forall\, x \in X$. The numbers are the membership degree $u_A(x)$ and $v_A(x)$ nonmembership degree of the element x to the set A, respectively.

Given an element x of X, the pair $(u_A(x), v_A(x))$ is called an intuitionistic fuzzy value (IFV) [6]. For ease of use, it can be denoted as $\tilde{a} = (u_{\tilde{a}}, v_{\tilde{a}})$ such that $u_{\tilde{a}} \in [0, 1]$, $v_{\tilde{a}} \in [0, 1]$, and $0 \leq u_{\tilde{a}} + v_{\tilde{a}} \leq 1$. The indeterminacy degree is denoted by $\pi_{\tilde{a}}$, with the conditions of $\pi_{\tilde{a}} \in [0, 1]$ and $\pi_{\tilde{a}} = 1 - u_{\tilde{a}} - v_{\tilde{a}}$.

In Definition 5.1, the membership and nonmembership degrees in a given universe are interpreted as points on the intuitionistic fuzzy interpretation triangle (IFIT). The geometrical depiction of IFS is shown in Fig. 5.1.

Definition 5.2 [20] Let $\tilde{a} = (u_{\tilde{a}}, v_{\tilde{a}})$ be an IFV, a score function S and an accuracy function H of the IFV $\tilde{a}$ is defined as the difference of membership and nonmembership function, as follows.

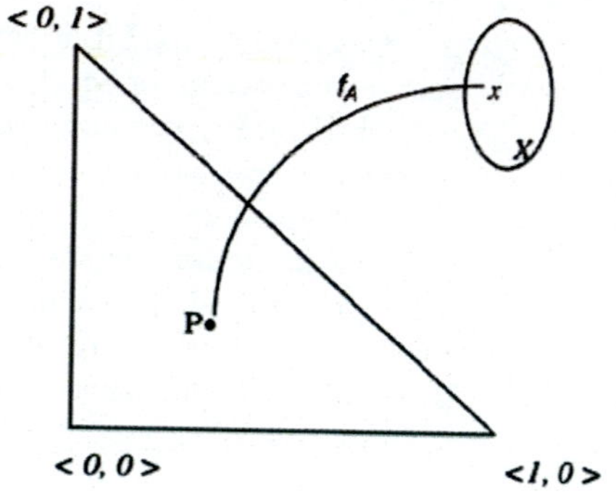

Fig. 5.1 Illustration of intuitionistic fuzzy set in IFIT [19]

$$S(\tilde{a}) = u_{\tilde{a}} - v_{\tilde{a}} \quad \text{where } S(\tilde{a}) \in [-1, 1] \tag{5.2}$$

$$H(\tilde{a}) = u_{\tilde{a}} + v_{\tilde{a}} \quad \text{where } H(\tilde{a}) \in [0, 1] \tag{5.3}$$

Definition 5.3 [19] Let $\tilde{a}_j = \left(u_{\tilde{a}_j}, v_{\tilde{a}_j}\right)$ be a set of IF pairs. Then, the IF weighted averaging (IFWA) operator is defined as follows.

$$\text{IFWA}_W(\tilde{a}_1, \tilde{a}_2, \ldots) = <1 - \prod\nolimits_{j=1}^{n} \left(1 - u_{\tilde{a}_j}\right)^{w_j}, \prod\nolimits_{j=1}^{n} v_{\tilde{a}_j}{}^{w_j}> \tag{5.4}$$

where $W = \{w_1, \ldots, w_n\}$ is the weighting vector of IF pairs with $w_j \in [0, 1]$ and $\sum_{j=1}^{n} w_j = 1$.

5.3 Methodology

The Selected Element Reduction Approach (SERA) is a relatively recent approach [4, 17] presented by the authors of this study for weighing criteria in a fuzzy environment, regardless of whether the criterion weight is obtained directly from a decision-maker or an expert. The influence of criteria on outcomes may also be used to evaluate their significance. The significance of the criteria is established by comparing the outcomes produced when a criterion is excluded. In a fuzzy environment, linguistic scales are helpful for the expert to express their opinions. The procedure aims to weight criteria based on the relationship of element reduction in multi-criteria decision-making. The proposed methodology can be adapted for any fuzzy sets.

The phases of the Intuitionistic Fuzzy Selected Element Reduction Approach (IF-SERA) are as follows:

Step 1: Define the case. Identify the sets of alternatives and criteria and decision-makers (DMs).

Step 2: Based on the criteria, collect intuitionistic fuzzy decision matrices from the DMs.

Step 3: Compute the combined fuzzy decision matrix of options using the IFWA aggregation operator in Eq. (5.4).

Step 4: Obtain aggregated fuzzy decisions of options using the IFWA aggregation operation in Eq. (5.4) (same as in Step 3). At this stage, the criteria are assumed to have equal weights.

Step 5: To determine the overall score ($S_{i_{OA}}$) of the alternatives, use score (a.k.a. defuzzification) function.

Step 6: Choose a criterion to reduce for each alternative, and calculate its score ($S_{i_{RC}}$). It means that the alternatives' scores are calculated without the reduced criterion.

Step 7: Apply the following equation to determine the effect of the reduction criteria.

$$E_{\mathrm{RC}_j} = \sum\nolimits_i \mid S_{i_{\mathrm{OA}}} - S_{i_{\mathrm{RC}_j}} \mid \tag{5.5}$$

Step 8: Assign the criteria weights by normalizing the reduced criterion's impact.

It should be noted that the aggregation operator and score (defuzzification) function varies depending on the kind of fuzzy collection. The flowchart in Fig. 5.2 depicts the IF-SERA.

5.4 Case Study

An evaluation of the IoT features of being a smart city is performed to demonstrate the application of the Intuitionistic Fuzzy Selected Element Reduction Approach (IF-SERA) for MCDM cases. In the light of relative studies in the literature, evaluation for four cities using four IoT criteria forms the basis of this study. Proposed IF-SERA methodology has been implemented gradually. Linguistic expressions are listed in Table 5.1 along with their intuitionistic fuzzy equivalents.

Step 1: A research is desired to evaluate the adaptations of four big cities $A = \{A_1, A_2, A_3, A_4\}$ in Turkey to IoT applications. In the study in which they will be compared, four criteria are determined as $C = \{C_1$: Energy Savings, C_2: Ensure Data Integrity and Quality, C_3 : Cloud Computing Integration, and C_4 : Administration and Coordination$\}$ [16], but their weights are not determined. Rather than taking the criteria subjectively from experts, weights specific to the entire evaluation are requested. In order to understand which criteria are more dominant in the evaluation, it is aimed to carry out a criterion weighting methodology. Intuitionistic fuzzy linguistic expressions are suitable for experts to evaluate the cities of member or nonmember of IoT features. The IF-SERA method, which is recent in the literature, is preferred to weight the criteria.

Step 2: Three decision-makers $DM = \{DM_1, DM_2, DM_3\}$ evaluated the alternative smart cities according to the criteria as follows, using the intuitionistic fuzzy scale given in Table 5.1.

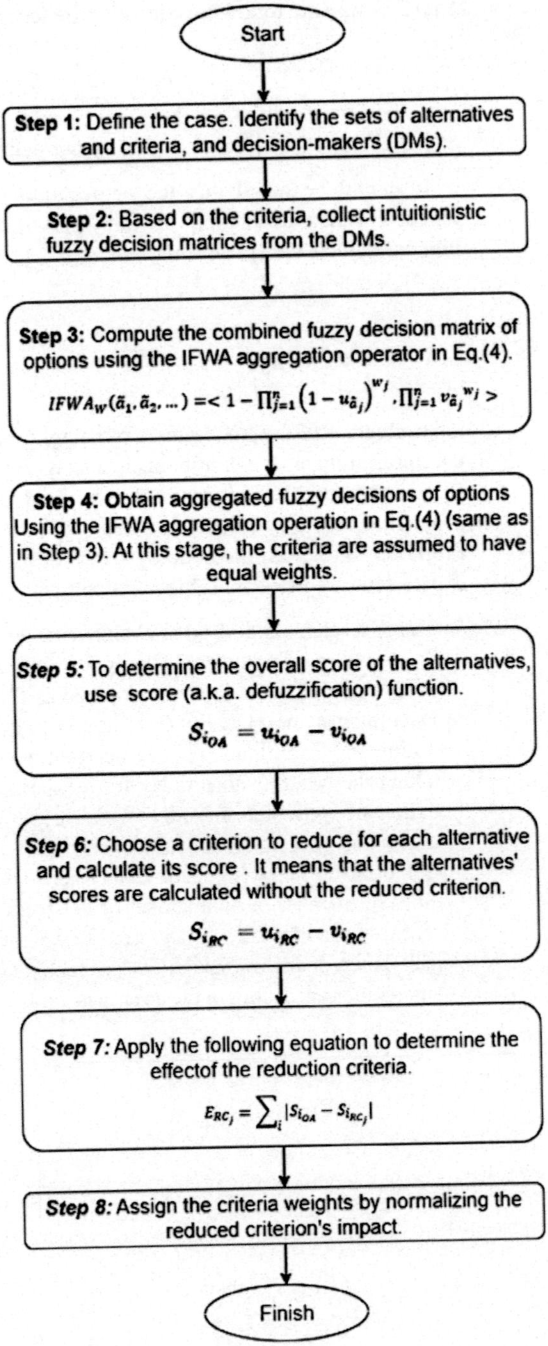

Fig. 5.2 The intuitionistic fuzzy selected element reduction approach flowchart in detail (IF-SERA) [17]

Table 5.1 Intuitionistic fuzzy linguistic expressions [17]

Scale	Intuitionistic fuzzy set	Scale	Intuitionistic fuzzy set
Extremely high (*EH*)	[1;0;0]	Medium low (*ML*)	[0.4;0.5;0.1]
Very high (*VH*)	[0.75;0.1;0.15]	Low (*L*)	[0.25;0.6;0.15]
High (*H*)	[0.6;0.25;0.15]	Very low (*VL*)	[0.1;0.75;0.15]
Medium high (*MH*)	[0.5;0.4;0.1]	Extremely low (*EL*)	[0.1;0.9;0]
Medium (*M*)	[0.5;0.5;0]		

$$DM_1 = \begin{matrix} & C_1 & C_2 & C_3 & C_4 \\ A_1 & EH & M & ML & MH \\ A_2 & H & VH & VH & ML \\ A_3 & VL & VH & M & H \\ A_4 & ML & M & ML & H \end{matrix}$$

$$DM_2 = \begin{matrix} & C_1 & C_2 & C_3 & C_4 \\ A_1 & MH & H & MH & L \\ A_2 & L & MH & M & M \\ A_3 & VL & MH & VH & ML \\ A_4 & M & M & VH & H \end{matrix}$$

$$DM_3 = \begin{matrix} & C_1 & C_2 & C_3 & C_4 \\ A_1 & H & MH & H & ML \\ A_2 & ML & H & ML & L \\ A_3 & EL & M & EH & M \\ A_4 & VL & VL & H & EH \end{matrix}$$

Step 3: The aggregated intuitionistic fuzzy decision matrix of alternatives by IFWA in Eq. (5.4) is calculated as follows:

$$DM_{\text{Aggr}} = \begin{matrix} & C_1 & C_2 & C_3 & C_4 \\ A_1 & (0.67, 0.23, 0.10) & (0.53, 0.39, 0.08) & (0.49, 0.39, 0.11) & (0.23, 0.61, 0.16) \\ A_2 & (0.49, 0.43, 0.08) & (0.61, 0.26, 0.13) & (0.53, 0.39, 0.08) & (0.37, 0.54, 0.10) \\ A_3 & (0.10, 0.82, 0.08) & (0.53, 0.39, 0.08) & (0.72, 0.23, 0.05) & (0.49, 0.43, 0.08) \\ A_4 & (0.27, 0.60, 0.13) & (0.29, 0.60, 0.10) & (0.56, 0.30, 0.13) & (0.71, 0.17, 0.11) \end{matrix}$$

Step 4: The aggregated intuitionistic fuzzy decisions of alternative cities obtained by using Eq. (5.4) are given as follows:

$$\tilde{D}_{\text{Aggr}} = \begin{matrix} A_1 \\ A_2 \\ A_3 \\ A_4 \end{matrix} \begin{bmatrix} (0.45,0.42,0.13) \\ (0.49,0.41,0.10) \\ (0.37,0.53,0.10) \\ (0.42,0.45,0.13) \end{bmatrix}$$

Step 5: The intuitionistic fuzzy numbers are defuzzified by Eq. (5.2). Thus, the overall score $(S_{i_{OA}})$ of alternatives is calculated as follows:

$$S_{i_{\text{OA}}} = \left\{ \begin{array}{c} S_{A1\text{OA}} = 0.026 \\ S_{A2\text{OA}} = 0.080 \\ S_{A3\text{OA}} = -0.158 \\ S_{A4\text{OA}} = -0.029 \end{array} \right\}$$

Step 6: Alternative scores are obtained by removing one element at a time from the evaluation. Return to Step 2, and search the matrices for the criteria column. The reevaluation criteria are now n-1. As the chosen criterion is reduced, the following alternative scores are calculated:

$$S_{i\text{RC}_1} = \left\{ \begin{array}{c} S_{A1\text{RC}_1} = -0.082 \\ S_{A2\text{RC}_1} = 0.085 \\ S_{A3\text{RC}_1} = 0.218 \\ S_{A4\text{RC}_1} = 0.101 \end{array} \right\} \quad S_{i\text{RC}_2} = \left\{ \begin{array}{c} S_{A1\text{RC}_2} = -0.008 \\ S_{A2\text{RC}_2} = 0.003 \\ S_{A3\text{RC}_2} = -0.239 \\ S_{A4\text{RC}_2} = -0.089 \end{array} \right\}$$

$$S_{i\text{RC}_3} = \left\{ \begin{array}{c} S_{A1\text{RC}_3} = 0.003 \\ S_{A2\text{RC}_3} = 0.061 \\ S_{A3\text{RC}_3} = -0.303 \\ S_{A4\text{RC}_3} = -0.110 \end{array} \right\} \quad S_{i\text{RC}_4} = \left\{ \begin{array}{c} S_{A1\text{RC}_4} = 0.217 \\ S_{A2\text{RC}_4} = 0.178 \\ S_{A3\text{RC}_4} = -0.222 \\ S_{A4\text{RC}_4} = -0.166 \end{array} \right\}$$

Step 7: The effects of the reduced criterion are given by Eq. (5.5) as follows:

$$E_{\text{RC}_j} = \left\{ \begin{array}{c} E_{\text{RC}_1} = 0.619 \\ E_{\text{RC}_2} = 0.309 \\ E_{\text{RC}_3} = 0.268 \\ E_{\text{RC}_4} = 0.489 \end{array} \right\}$$

Step 8: The weights of criterion are calculated by normalizing the criterion effects as follows:

$$w_j = \left\{ \begin{array}{c} w_{C_1} = 0.367 \\ w_{C_2} = 0.184 \\ w_{C_3} = 0.159 \\ w_{C_4} = 0.290 \end{array} \right\}$$

The findings show that $C1$: *Energy Savings* is the most important criterion for assessing the adaptation of cities to IoT features.

5.5 Conclusion

This study investigated the adaptation of smart cities to IoT technologies through the proposed IF-SERA procedure. In this methodology, the effect of successively reducing the criteria selected from the process is examined, and this effect ratio is evaluated as the criteria weight in an intuitionistic fuzzy environment without subjective weighting. Smart cities are those that use the smart devices to do a variety of tasks, including energy consumption, pollution reduction, traffic management, lighting, and traffic control. The ability of cities to adapt to these features with smart devices increases the quality of life. Therefore, IF-SERA method is performed to evaluate the adaptation of cities to smart life and is applied based on "Ensure Data Integrity and Quality, Administration and Coordination, Energy Savings Cloud, and Computing Integration" criteria.

For more benefits, different fuzzy sets can be integrated into the SERA procedure, and it is recommended to expand its use by applying it to different MCDM fields in the literature. In addition, it has been suggested that SERA can be applied as a hybrid with MCDM methods such as AHP, TOPSIS, VIKOR, EDAS, etc. whose effectiveness has been proven in the literature.

Acknowledgments "This work has been supported by the Scientific Research Projects Commission of Galatasaray University under grant number # FBA-2022-1085."

References

1. B. Hammi, R. Khatoun, S. Zeadally, A. Fayad, L. Khoukhi, IoT technologies for smart cities. IET Netw. **7**(1), 1–13 (2018)
2. T. Alam, Cloud-based IoT applications and their roles in smart cities. Smart Cities **4**(3), 1196–1219 (2021)
3. P. Bellini, P. Nesi, G. Pantaleo, IoT-enabled smart cities: A review of concepts, frameworks and key technologies. Appl. Sci. **12**(3), 1607 (2022)
4. E. Çakır, M. Taş, E. Demircioğlu, A new weighting method in fuzzy multi-criteria decision making: Selected element reduction approach (SERA), in *North American Fuzzy Information Processing Society Annual Conference*, (Springer, Cham, 2022), pp. 20–30
5. L.A. Zadeh, Fuzzy sets. Inf. Control. **8**, 338–356 (1965)

6. K. Atanassov, Intuitionistic fuzzy sets. Fuzzy Sets Syst. **20**, 87–96 (1986)
7. R. Sadiq, S. Tesfamariam, Environmental decision-making under uncertainty using intuitionistic fuzzy analytic hierarchy process (IF-AHP). Stoch. Env. Res. Risk A. **23**(1), 75–91 (2009)
8. J.Q. Wang, R.R. Nie, H.Y. Zhang, X.H. Chen, Intuitionistic fuzzy multi-criteria decision-making method based on evidential reasoning. Appl. Soft Comput. **13**(4), 1823–1831 (2013)
9. F. Shen, X. Ma, Z. Li, Z. Xu, D. Cai, An extended intuitionistic fuzzy TOPSIS method based on a new distance measure with an application to credit risk evaluation. Inf. Sci. **428**, 105–119 (2018)
10. A.R. Mishra, P. Rani, K. Pandey, A. Mardani, J. Streimikis, D. Streimikiene, M. Alrasheedi, Novel multi-criteria intuitionistic fuzzy SWARA–COPRAS approach for sustainability evaluation of the bioenergy production process. Sustainability **12**(10), 4155 (2020)
11. Ş. Şeker, IoT based sustainable smart waste management system evaluation using MCDM model under interval-valued q-rung orthopair fuzzy environment. Technol. Soc. **71**, 102100 (2022)
12. G. Özkaya, C. Erdin, Evaluation of smart and sustainable cities through a hybrid MCDM approach based on ANP and TOPSIS technique. Heliyon **6**(10), e05052 (2020)
13. H. Mohapatra, B.K. Mohanta, M.R. Nikoo, M. Daneshmand, A.H. Gandomi, MCDM based routing for IoT enabled smart water distribution network. IEEE Internet Things J. **10**(5), 4271–4280 (2022)
14. M. Lin, C. Huang, Z. Xu, R. Chen, Evaluating IoT platforms using integrated probabilistic linguistic MCDM method. IEEE Internet Things J. **7**(11), 11195–11208 (2020)
15. G. Büyüközkan, D. Uztürk, Smart last mile delivery solution selection for cities, in *Proceedings of the World Congress on Engineering*, (2019)
16. H. Rajab, T. Cinkelr, IoT based smart cities, in *2018 International Symposium on Network Computers and Communications (ISNCC)*, 1–4 (2018)
17. E. Çakır, M. Taş, E. Demircioğlu, Intuitionistic fuzzy selected element reduction approach (IF-SERA) on service quality evaluation of digital suppliers, in *International Conference on Information Processing and Management of Uncertainty in Knowledge-Based Systems*, (Springer, Cham, 2022), pp. 141–150
18. S. Chen, J. Tan, Handling multicriteria fuzzy decisionmaking problems based on vague set theory. Fuzzy Sets Syst. **67**(2), 163–172 (1994)
19. A.G. Hatzimichailidis, B.K. Papadopoulos, A new geometrical interpretation of some concepts in the intuitionistic fuzzy logics. NIFS **11**(2), 38–46 (2005)
20. Z. Xu, Intuitionistic fuzzy aggregation operators. IEEE Trans. Fuzzy Syst. **15**(6), 1179–1187 (2007)

Chapter 6
Predicting Cyber-Trafficking Websites Using a Naive Bayes Algorithm, Logistic Regression, KNN, and SVM

Aiza Jane Sulit, Risty Acerado, Ramon Christus Tomaquin, and Roselia Morco

Abstract Researchers and software developers continue discovering the best approach to combat the rising cyber-trafficking issues. However, most studies focus only on one platform or one of the gateways of cyber-trafficking. Thus, this paper introduces the development and comparison of the Naive Bayes Algorithm, Logistic Regression, k-nearest neighbor (KNN), and Support Vector Machine (SVM) classification models to predict trafficking and non-trafficking websites. In developing the supervised classification models, 37 keywords were used to scrape suspected trafficking websites. Thirty-five (35) websites were classified as trafficker out of 63; this data was used to create the models. Upon evaluating the accuracy rates of the models, the Naive Bayes Algorithm got ninety-one percent (91%), Logistic Regression got eighty-one percent (81%), KNN got sixty-four percent (64%), and SVM got sixty-four percent (64%). Thus, Naive Bayes can predict more accurately than the other classification algorithms. The result shows that the predictive model could be an effective tool for identifying different online platforms that are used in trafficking. Once the model is integrated into an application, this will be easier and faster for law enforcement agencies to monitor human trafficking in a fast-growing cyberspace community.

Keywords Naïve bayes · Logistic regression · K-nearest neighbor (KNN) · Support vector machine · Cyber trafficking · Human trafficking · Classification algorithm

A. J. Sulit · R. Acerado (✉) · R. C. Tomaquin · R. Morco
Information Systems Department Technological Institute of the Philippines, Quezon City, Philippines
e-mail: racerado.is@tip.edu.ph

C.-C. Wang, A. Nallanathan (eds.), *6th International Conference on Signal Processing and Information Communications*, Signals and Communication Technology, https://doi.org/10.1007/978-3-031-43781-6_6

6.1 Introduction

Human trafficking cases have been increasing continuously. Human traffickers target no exceptions of age, gender, and race. However, most reported victims worldwide are mostly the vulnerable ones, women and girls, and they are mostly bound for sexual exploitation [1, 2]. In the Philippines, commercial sex exploitation usually occurs near offshore gaming operations and tourist destinations [3]. But now, sex trafficking is also prevalent in cyberspace, leading to many cybersex trafficking cases. The popularization and free access to many online platforms, such as social media and video-sharing sites, have opened many opportunities to traffickers. These online activities in chat rooms, social networking sites, online ads, and many other social media sites have enabled traffickers to target more victims [4].

As cyber trafficking is unstoppable, the data related to these cases are growing. However, there is no assurance that all the data stored in the government's database are "good" data as, according to [5], the data management process is poor. The challenge of analyzing good data paves the way for developing tools for information extraction, data mining, and machine learning. Experts can use these techniques to identify patterns with pertinent information on human trafficking on the internet [6]. In addition, many terminologies are used to identify human trafficking activities. For example, the terms "pimp" or "madam" are employed to indicate the probable trafficker in the situation, a "provider," who is the same as the person being sold. The word "johns" is also a known word referring to the customers of this online trafficking business [7]. Realizing the greater challenges faced by the government and seeing the opportunities to help, many researchers were interested in analyzing the activities and strategies of traffickers. Thus, many research studies were conducted using advanced technologies and innovations such as sentiment analysis [8–10] and natural language processing [11]. Most studies just focused on one social platform, such as social media messages [12], dark web [13], website advertisements [14, 15], open internet sources [16], and Twitter posts [17].

The data quality and the learning algorithms' efficacy all play a role in determining how successful a machine-learning solution will be [18]. Therefore, researching numerous machine learning algorithms enables one to determine which algorithms may be combined to provide the most accurate predictive model for monitoring websites. Studies that present classification predictions of cyber-trafficking websites are still limited. Most studies focus only on one platform or one of the gateways of cyber-trafficking. Therefore, this paper introduces the development and comparison of the Naive Bayes Algorithm, Logistic Regression, KNN, and SVM classification models to predict trafficking and non-trafficking websites.

6.2 Review of Related Studies and Literature

This section presents reviews of literature and studies related to trafficking and analytics.

6.2.1 Review on the Human Trafficking Cases

The data for the visualizations in this section was derived from the Counter-Trafficking Data Collaborative (CTDC), the world's first global data hub on human trafficking, which publishes standardized data from anti-trafficking groups worldwide [19, 20].

Figure 6.1 shows that not only the most vulnerable in the society are being targeted by traffickers.

Human trafficking comes from different forms. Trafficking is really active in online and offline global trading. Figure 6.2 shows that sexual exploitation is the major market for human traffickers. It reflects that traffickers gain more profit in this form of trafficking.

Human trafficking is not bounded by time and space, as presented in Fig. 6.3. There are many instances around the globe. Unfortunately, the Philippines has the highest proportion of trafficking occurrences among all nations, as shown in Fig. 6.3, with 11,365 instances, followed by Ukraine, which has 7,761 instances.

Figure 6.4 indicates that the vast majority of victims did not provide any specific information about their interaction with the traffickers. Unfortunately, even the victims' relatives are the primary reason victims suffer in misery in exchange for monetary compensation.

Fig. 6.1 Pie chart distribution of gender

Fig. 6.2 Tabulation of the different forms of human trafficking

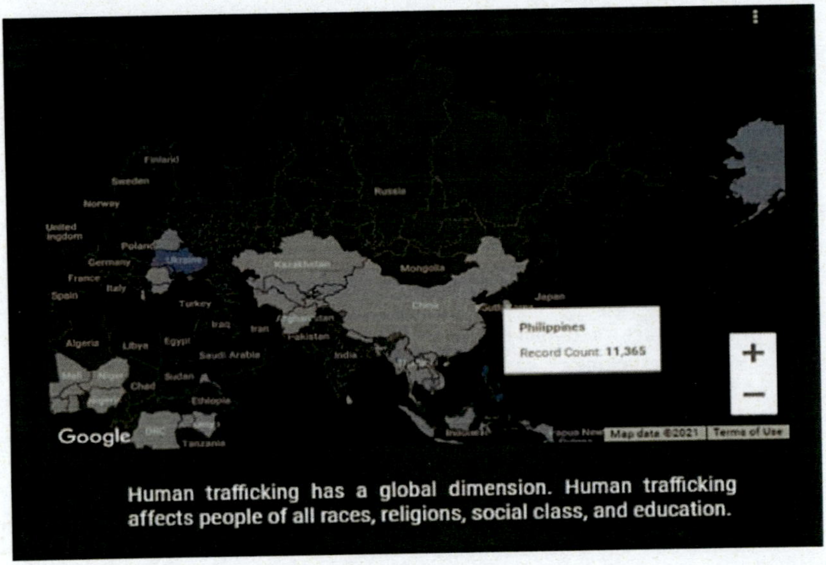

Fig. 6.3 Filled map chart for global trafficking cases

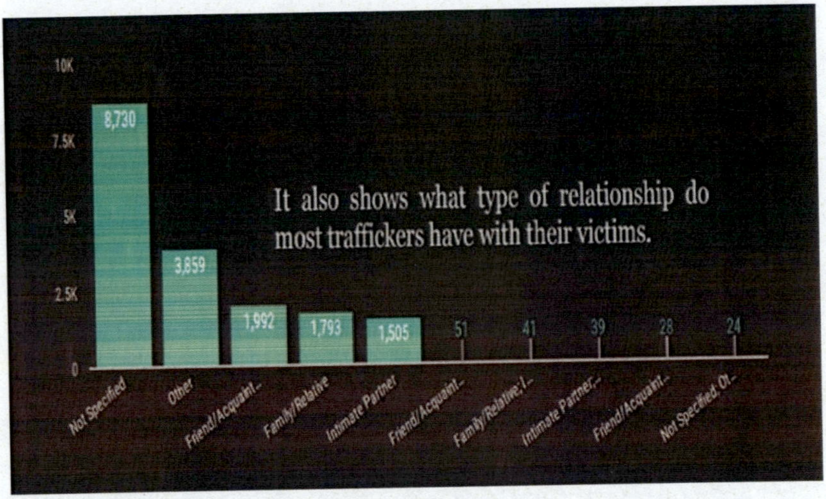

Fig. 6.4 Bar chart showing the recruitment relationship

6.2.2 Review on the Analytics to Combat Human Trafficking

Predictive analytics is the branch of advanced analytics used to predict unknown future events. This study used predictive analytics to determine a potential trafficking website using predetermined keywords. Like in the concept of Search Engine Optimization (SEO), keywords can be used to predict top search results. The study of [21] has integrated it with machine learning to predict SEO rankings. The same approach was applied in this study that a set of keywords were identified to determine a website that contains keywords most probably used by online traffickers.

Naive Bayes shows satisfactory results even in small-scale datasets [22]. Therefore, in this study, Naive Bayes was also used. In addition, Naive Bayes classifiers have better resilience to missing data than support vector machines [20]. Additionally, other machine learning algorithms, logistic regression, KNN, and SVM were also employed to evaluate their performance in predicting cyber trafficking.

6.3 Development of Cyber Trafficking Websites Classification Models

Naive Bayes classifier, Logistic Regression, KNN, and SVM were used to develop the model for classifying cyber-trafficking websites. In addition, it was used to forecast traffic on both trafficking and non-trafficking websites.

The data was extracted from scraped websites using a scraping algorithm written in the Python programming language and 37 keywords marked as red flags for traffickers. The websites were categorized into two types: trafficker and non-trafficker. The categorization process was done with the help of an expert who visited the websites with special tools. However, due to project time constraints, there were only 35 websites that were classified as trafficker out of the 63 total scraped websites. The data that was collected was formatted into a spreadsheet.

Then, to manage and analyze the data used for the analysis, the data were first cleaned by removing some of the less significant data and converting a portion of the data from text to numeric, since most of the data is string type. Encoding of string-type data is conducted so that the machine learning algorithm can execute its arithmetic operation to understand the data that must be analyzed. The algorithms performed for this study do not accept string values. Only those that are close to floating data types, which is why a label encoder is used. The data was separated into two categories: features and encodedClass. The feature category contains type and keyword data, whereas the encoded class category contains all other data. The following elaborates on the columns of the dataset:

- Url is the link to the website.
- Type indicates the type of the website or its general description such as a news article, eCommerce, educational, nonprofit, blogs, portfolio, portal, and search engines/job search engines.
- Keywords are one of the 37 keywords with the highest occurrences on the particular website.
- Keycounts corresponds to the number of occurrences of the keyword appearing on a website.
- Date is the date that the website was posted. When no date indicates the website or page that contains the keywords, the date was set to the date of the data collection.

The accuracy of models created using Naive Bayes Algorithm, Logistic Regression, KNN, and SVM was compared to determine know which is best suited for the data on hand. The algorithm examines and calculates occurrences of keywords found on trafficking or non-trafficking websites.

6.4 Evaluation of the Model

Using the train-test split library in the sklearn package and the 80/20 ratio of the training and validation data sets, the dataset is split to produce the training and testing sets. The evaluation yields two classes: class 0 for non-traffickers and class 1 for traffickers.

Figure 6.5 shows the classification report of the logistic regression model. Out of all the non-traffickers, the model predicted that 100% of them were correctly classified. Similarly, classifying traffickers also accumulated a 100% correct classification rate. The value result showed that the model does a great job of predicting whether or not it is a trafficking or non-trafficking website.

Precision and recall indicate a missing value in the accuracy column. This is because data accuracy should be as high as possible, and comparing the two models becomes difficult if the data has a low precision but a high recall, or vice versa. Therefore, to evaluate the results of the accuracy test, the F-score is used to evaluate both the precision and the recall of the data.

Fig. 6.5 Classification report for logistic regression

	precision	recall	f1-score	support
0	1.00	1.00	1.00	3
1	1.00	1.00	1.00	4
accuracy			1.00	7
macro avg	1.00	1.00	1.00	7
weighted avg	1.00	1.00	1.00	7

	precision	recall	f1-score	support
0	1.00	1.00	1.00	3
1	1.00	1.00	1.00	4
accuracy			1.00	7
macro avg	1.00	1.00	1.00	7
weighted avg	1.00	1.00	1.00	7

KNN

Fig. 6.6 Classification report for k-nearest neighbors

	precision	recall	f1-score	support
0	1.00	1.00	1.00	3
1	1.00	1.00	1.00	4
accuracy			1.00	7
macro avg	1.00	1.00	1.00	7
weighted avg	1.00	1.00	1.00	7

Fig. 6.7 Classification report for Naive Bayes

	precision	recall	f1-score	support
0	0.75	1.00	0.86	3
1	1.00	0.75	0.86	4
accuracy			0.86	7
macro avg	0.88	0.88	0.86	7
weighted avg	0.89	0.86	0.86	7

Fig. 6.8 Classification report for SVM

Figure 6.6 shows the classification report of the K-Nearest Neighbors model. The model showed a similar output to the logistic regression model with 100% correctly identified positives as well as the same for the accuracy of each positive prediction, while the support showed a balanced dataset consisting of three (3) non-traffickers and four (4) traffickers.

Figure 6.7 shows that in the same manner as with the K-nearest neighbors and logistic regression, Naive Bayes model has 100% correctly identified positives. The same goes for the accuracy of each positive prediction as well as the support.

Figure 6.8 shows the classification report of the SVM model. The model predicted that out of all the non-traffickers, 75% of them were correctly classified. On the other hand, classifying traffickers accumulated a 100% correct classification

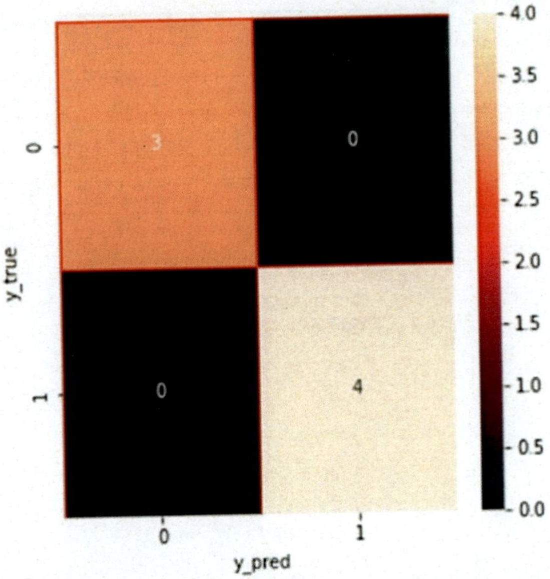

Fig. 6.9 Confusion matrix for logistic regression, K-nearest neighbors (KNN), and Naive Bayes

rate. The F1 score value appears to be near 1, indicating that the model is good at predicting both non-traffickers and traffickers while also maintaining support at an acceptable range.

Figure 6.9 shows the confusion matrix for logistic regression, SVM, and Naive Bayes. These three models appeared to share a classification report, resulting in the same confusion matrix. The confusion matrix shows no misclassifications that appear in the model. Instead, the confusion matrix demonstrates that the classification report's three non-trafficker websites and four trafficker websites appeared to be correctly classified. There are seven correctly classified websites for these models.

A confusion matrix is utilized because it allows visitors to examine the outcomes of an algorithm at a glance. The confusion matrix presents analytical findings in the form of a straightforward table, which effectively condenses the outputs into a perspective that is easier to understand.

As shown in Fig. 6.10, there are six misclassifications. Three on the non-trafficking website and three on the trafficking website. The correctly classified website is only one (1).

Figure 6.11 shows the result of the cross-validations for all the models presented. The KFold validation was used in determining the accuracy of the data in all the models. This shows that although three of the models have the same classification report, they still differ when it comes to the accuracy of their data. Cross-validation is conducted to get more information about the algorithm performance. The validation showed that Naive Bayes appears to be the highest among all the models.

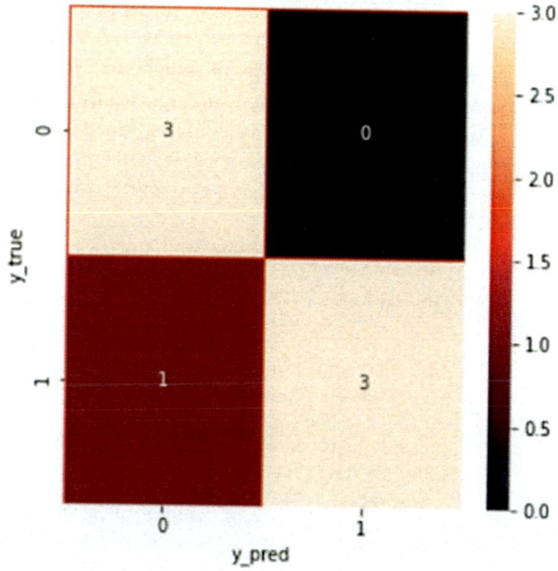

Fig. 6.10 Confusion matrix for SVM

Fig. 6.11 Cross-validation results

```
Logistic Regression:  0.8181818181818182
SVM:  0.6363636363636364
KNN:  0.6363636363636364
Naive Bayes:  0.9090909090909091
```

6.5 Conclusion

The presented predictive algorithm using four machine learning algorithms, namely, the Naive Bayes Algorithm, Logistic Regression, K-Nearest Neighbor (KNN), and Support Vector Machine (SVM), was able to predict trafficking and non-trafficking websites. The evaluation shows that the model performs well, having an accuracy result of 91%, 81%, 64%, and 64%, respectively, in classifying trafficking and non-trafficking websites. Furthermore, as observed from the classification report, some machine learning algorithms reflect the same value, because it reflects the measurement's proximity to the actual value, but they are not identical. Accuracy indicates how close a measurement is to a known or accepted value, regardless of how far it deviates from the accepted value. Both precise and accurate measurements are repeatable and close to the true values.

Moreover, Naive Bayes stands out with the highest accuracy rate, making it an ideal model to be used, also taking into consideration the results from SVM, which is acceptable enough to be used along with the Naive Bayes, which could garner a more effective outcome for predictions. The model can be integrated into a tool to help law enforcement agencies to analyze transactions on the web and identify possible cyber traffickers. It is an excellent line of defense to act on early signs of human trafficking before it has taken a victim. In this way, law enforcement can develop a task force that will watch over red flags identified by the analytical models presented. Moreover, scrutinizing each online transaction would be much easier for law enforcement agencies, allowing them to act on the concerns of a larger community in need of their assistance. It shows that the model can be helpful in proactively combating traffickers that are roaming around in cyberspace.

References

1. Stop The Traffik. Definition and Scale.Learn more about human trafficking and how prevalent this fast-growing crime is. 2022. Retrieved in February, 2023. https://www.stopthetraffik.org/what-is-human-trafficking/definition-and-scale/
2. United Nations Office on Drugs and Crime (UNODC). Global report on trafficking in persons. 2009. Retrieved in February 2023. https://www.unodc.org/unodc/en/press/releases/2023/January/global-report-on-trafficking-in-persons-2022.html
3. U.S. Department of State. Trafficking in persons report: Philippines. 2022. Retrieved in February 2023. https://www.state.gov/reports/2022-trafficking-in-persons-report/philippines__trashed/
4. M. Latonero. Human trafficking online: the role of social networking sites and online classifieds. Center on Communication Leadership & Policy. Research Series: September 2011. USC. Annenberg School for communication & Journalism https://doi.org/10.2139/ssrn.2045851
5. J. Brunner. Getting to good human trafficking data. Assessing the landscape in Southeast Asia and promising practices from ASEAN governments and civil society. 2018. Retrieved in February 2023. https://humanrights.stanford.edu/sites/humanrights/files/good_ht_data_policy_report_final.pdf
6. D. Burbano, M. Hernandez-Alvarez. Identifying human trafficking patterns online, in IEEE Second Ecuador Technical Chapters Meeting (ETCM) (2017), pp. 1–6, doi: https://doi.org/10.1109/ETCM.2017.8247461
7. M. Ibanez, D. D. Suthers. Detection of domestic human trafficking indicators and movement trends using content available on open internet sources, in 47th Hawaii International Conference on System Sciences, pp. 1556–1565, 1, 2014
8. K.S.A. Rose, Application of sentiment analysis in web data analytics. Int. Res. J. Eng. Technol. **07**(06), 2395 (2020)
9. D. Liu, C.Y. Suen, O. Ormandjieva. A novel way of identifying cyber predators. ArXiv, abs/1712.03903 (2017)
10. A. Mensikova, C.A. Mattmann Ensemble sentiment analysis to identify human trafficking in web data, in Workshop on Graph Techniques for Adversarial Activity Analytics (GTA). Marina Del Rey, 2018, pp. 5–9
11. Maria Diaz, Anand V. Panangadan. Natural language-based integration of online review datasets for identification of sex trafficking businesses, in IEEE 21st International Conference on Information Reuse and Integration for Data Science (IRI) (2020), pp. 259–264
12. W. Chung, E. Mustaine, D. Zeng. Criminal intelligence surveillance and monitoring on social media: Cases of cyber-trafficking, in 2017 IEEE International Conference on Intelligence and Security Informatics (ISI). IEEE Press (2017), pp. 191–193. https://doi.org/10.1109/ISI.2017.8004908
13. C.A. Murty, P.H. Rughani, Sentiment & pattern analysis for identifying nature of the content hosted in the dark web. Indian J. Comput Sci Eng **12**(6) (2021). https://doi.org/10.21817/indjcse/2021/v12i6/211206142

14. A. Volodko, E. Cockbain, B. Kleinberg, "Spotting the signs" of trafficking recruitment online: Exploring the characteristics of advertisements targeted at migrant job-seeker. Trends Organ Crim **23**, 7–35 (2020). https://doi.org/10.1007/s12117-019-09376-5
15. M. Latonero. Human trafficking online: The role of social networking sites and online classifieds. Center on Communication Leadership & Policy. Research Series: September 2011. USC. Annenberg School for communication & Journalism https://doi.org/10.2139/ssrn.2045851
16. H. Wang, C. Cai, A. Philpot, M. Latonero, E.H. Hovy, D. Metzler. Data integration from open internet sources to combat sex trafficking of minors, in Proceedings of the 13th Annual International Conference on Digital Government Research (2012), pp. 246–252. https://doi.org/10.1145/2307729.2307769
17. M.H. Alvarez, D.B. Acuña. (2017). Identifying human trafficking patterns online, in Conference: 2017 IEEE Second Ecuador Technical Chapters Meeting (ETCM), Salinas Ecuador (2017). DOI:https://doi.org/10.1109/ETCM.2017.8247461
18. I.H. Sarker. *Machine Learning: Algorithms, Real-World Applications and Research Directions – SN Computer Science*. SpringerLink. (2021). Retrieved May 29, 2022 from https://link.springer.com/article/10.1007/s42979-021-00592-x
19. Counter-Trafficking Data Collaborative. https://respect.international/counter-trafficking-data-collaborative/
20. Hongbo Shi, Yaqin Liu. Naïve Bayes vs. Support vector machine: resilience to missing data. Lecture Notes in Computer Science (2011), p. 8 https://doi.org/10.1007/978-3-642-23887-1_86
21. Michael Weber. Machine learning for SEO – How to predict rankings with machine learning (October 26, 2017). Retrieved June 2021 from https://www.searchviu.com/en/machine-learning-seo predicting-rankings/
22. Yuguang Huang, Lei Li. Naive Bayes classification algorithm based on a small sample set, in IEEE International Conference on CloudComputing and Intelligence Systems (September 2011), p. 6. DOI:https://doi.org/10.1109/CCIS.2011.6045027

Chapter 7
Flood Forecasting Using Edge AI and LoRa Mesh Network

Mau-Luen Tham, Xin Hao Ng, Rong-Chuan Leong, and Yasunori Owada

Abstract Remote flood forecasting has exponentially grown over the past decade together with the unprecedented expansion of Internet of Things (IoT) network. This is feasible with the use of long-range wireless communication technology such as LoRa. Ideally, each LoRa device shall process the sensor data locally and trigger warnings to the remote server based on prediction results. However, conventional prediction methods rely on highly computational artificial intelligence (AI) algorithms, which are not suitable for low-powered LoRa network. In this paper, the LoRa device is integrated with an edge AI model, which is based on long short-term memory (LSTM) neural network. OpenVINO is adopted to optimize the LSTM model before executing the solution on a Raspberry Pi 4 in combination with Intel Movidius Neural Computing Stick 2 (NCS2). Experimental results demonstrate the feasibility of deployment of the customized model on low-cost and power-efficient embedded hardware.

Keywords Edge AI · LSTM · Flood forecasting · LoRa Mesh Network · IoT

7.1 Introduction

Flood forecasting models have been researched in the hydrological engineering area for many years. Recently, there has been increased research interest in river flood prediction and modeling, defined as data-driven approaches. The artificial neural network (ANN) model is the most famous usual data-driven approach. Most

M.-L. Tham (✉) · X. H. Ng · R.-C. Leong
Department of Electrical and Electronic Engineering, Universiti Tunku Abdul Rahman, Kajang, Malaysia
e-mail: thamml@utar.edu.my; lrongchuan@1utar.my

Y. Owada
Resilient ICT Research Center, National Institute of Information and Communications Technology (NICT), Tokyo, Japan
e-mail: yowada@nict.go.jp

C.-C. Wang, A. Nallanathan (eds.), *6th International Conference on Signal Processing and Information Communications*, Signals and Communication Technology, https://doi.org/10.1007/978-3-031-43781-6_7

conventional statistical methods require a lot of data for their models, and they can generate no assumptions for both linear and nonlinear systems. Hence, the data-driven approach, ANN, is an alternative to hydrological flood forecasting instead of the existing methods [1].

Artificial intelligence (AI) has made essential development in modeling hydrological forecasting and dynamic hydrological issues. With the advancement of information technology, the application of ANN models in many aspects of science and engineering is increasingly becoming common due to its simplicity of structure. Diverse neural network modeling approaches have been applied, like implementing the model approaches individually or combining process-based approaches to minimize mistakes and increase the models' forecasting accuracy. The study in [2] applied AI model to forecast river flow for 15 years starting from 2000.

However, there are some limitations of the ANN model. One of them is lacking understanding of watershed processes. Furthermore, the limitation of memory in calculating sequential data exposes the disadvantages of the ANN model. The breakthrough in computational science has recently increased the interest in deep neural network (DNN) approaches. In addition, the most recent DNN applications, such as the long short-term memory (LSTM) [3] and gated recurrent unit (GRU) [4] neural networks, have been efficiently implemented in diverse areas and fields, such as time sequence problems. Those models can apply to machine translation, speech recognition, tourism field, language modeling, rainfall-runoff simulation, stock prediction, and river flow forecasting.

On 11th March 2011, around 29,000 cellular towers were damaged in the East Japan Great Earthquake. These damages have restricted the broadcast of evacuation notices and the collection of historical information for disaster forecasting. Hence, it can be known that the resilience of a network remains an open issue in the deployment of the fault-tolerant network during an emergency disaster. Fortunately, a disaster-resilient mesh-topological network called NerveNet was developed by Japan NICT. Each NerveNet node is independent and tolerant to system failure and link disconnection due to its mesh structure.

In this paper, a flood forecasting model is proposed. In the study area, rainfall and river water levels collected at hydrological stations serve as dataset for the training and testing process of the AI models. Then, the forecasted flood water level will be processed to generate the flood warning message. It will be sent through the NerveNet LoRa mesh network. Note that the proposed solution facilitates edge computing, which is one of goals of the ASEAN IVO project titled "Context-Aware Disaster Mitigation using Mobile Edge Computing and Wireless Mesh Network."

The rest of the paper is organized as follows. Section II discusses the related works. Section III describes the system architecture. Section IV presents the experimental results and discussions. Section V concludes the article.

7.2 Related Work

7.2.1 Edge AI

Several existing works [5, 6] explored the potential of edge AI for various applications. The authors in [5] focused on real-time apple detection with the implementation of YOLOv3-tiny algorithm on various embedded platforms. However, they did not consider the communication aspects. Recognizing the importance of LoRa, the authors in [6] proposed an edge AI in LoRa-based fall detection system with fog computing and LSTM. The processing burden is placed on a LoRa-based edge gateway, where the collected sensor information is transmitted from an edge node via Bluetooth Low Energy (BLE). Differently, our solution integrates both edge AI and LoRa functionalities into one single device, which simplifies the deployment effort.

7.2.2 NerveNet

NerveNet is a resilient network developed by Japan National Institute of Information and Communications Technology (NICT) [7]. NerveNet is a specially developed network for the regional area to provide reliable network access and a stable, resilient information-sharing platform in emergencies, even if the base station is destroyed in a disaster. The base stations of NerveNet are interconnected by the Ethernet-based wired or wireless transmission systems such as satellite, Wi-Fi, LoRa, and so on. They will form a mesh-topological network.

Nowadays, the current trend of the common network infrastructures uses the tree topology. As compared to it, NerveNet has the characteristic that it is more tolerant to the faults such as node failures, disconnections, destruction of the base station, and so on. Since the base station in the NerveNet supports basic services such as SIP proxy, DNS, and DHCP, the NerveNet can also continuously provide connectivity services to the devices.

7.3 System Architecture

7.3.1 Dataset

The dataset we employ is the Abashiri River watershed [8], located northeast of Hokkaido, Japan. The area of the watershed is around 1380 km^2. It has a 115 km long main river to the North Pacific and a range of elevation from 0 to 978 m [9]. All AI models are trained and tested using the datasets observed at the downstream

Table 7.1 Training and testing period for the dataset

Dataset	Training	Test
Hongou (Jan 2019 to Dec 2020)	**Jan 2019 to May 2020**	**Jun 2020 to Dec 2020**

Table 7.2 Hyperparameter settings for LSTM model

Hyperparameter	Value
Sequence length	**24**
Optimization algorithm	**Root mean squared propagation**
Epoch	**50**
Batch size	**64**

stations called "Hongou." The used datasets are hourly datasets with the water level and rainfall variables from first January 2019 to 31st December 2020.

During data preprocessing, the rainfall and water level data undergo a train-test split, separated into 70% of the data as training dataset and 30% as a testing dataset, as listed in Table 7.1. The training data calculates the training process error and estimates the AI models' parameters. The testing data provides an independent performance evaluation of the AI models after training.

Next, the hydrological dataset has also undergone data standardization where the values' distribution is rescaled to a mean value of 0 and a standard deviation value of 1. Data scaling is essential to fasten the training process of the AI model because the AI models can converge more rapidly if the dataset features are closer to the normal distribution. Prior to the AI model training, the time series dataset is converted into sequential data with 24-time steps as the sequence length. The model performs equally well when the sequence length is between 5 and 15 or more. Therefore, in this paper, the sequence length value of 24 is used in the model to represent 24 h in 1 day.

7.3.2 *AI Model Training in Google Colab*

In this paper, four types of AI models, namely, Random Forest, SVM, LSTM, and GRU, are trained and tested on the dataset to benchmark the performance of the system in terms of flood water level forecasting. Trained in in Google Colab platform, the best AI model will be selected as the edge AI.

For Random Forest, the parameter "max_depth" represents each tree's depth in the forest. Here, we set the max_depth value to 2. There are several hyperparameters in the LSTM model-building process. Firstly, the optimization algorithm is the stochastic gradient descent procedure's extension to update the weights iterative of the network according to the training dataset. Secondly, an epoch is defined as the whole dataset transferring forward and backward across the model's neural network once. Thirdly, the batch size is the number of samples propagating throughout the entire neural network. Table 7.2 demonstrates the hyperparameter settings of the

LSTM model. For fair comparison, the same hyperparameters are adopted to train the GRU models.

7.3.3 AI Model Optimization Using OpenVINO

The immediate output format of the LSTM model is .h5, which will be converted to pb format. The intention is to utilize the OpenVINO toolkit [10], which enables the faster running of the AI model in edge device. There are two main components in the OpenVINO toolkit, which are the model optimizer and inference engine. Firstly, when the trained model in pb format is fed into the model optimizer, it converts them to the IR format. At the same time, it optimizes the performance, space, and hardware-agnostic with conservative topology transformations. The outputs of the model optimizer are .xml and .bin.

Secondly, the AI inferencing process is performed at the edge device by setting the inference engine to Intel Neural Compute Stick 2 (NCS2), which is a hardware accelerator. Before feeding to the inference engine, the data is scaled using the scaler.gz exported from the training process. The scaled data is then reframed. The historical time series data representing the last 24 h is extracted from the scaled dataset by retrieving the top 24 values of the rainfall and water level data. After that, the sequence data and the trained model in IR format are fed into the inference engine to generate the water levels ahead of 1 hour in text form and the result graph in image form.

7.3.4 Evaluation Metrics

The mean absolute error (MAE) is the mean of the differences between the original value with the forecasted value. On an excellent flood forecast, the MAE should be smaller. Mathematically, it can be expressed as:

$$\mathrm{MAE} = \frac{1}{n}\sum_{i=1}^{n}|e_i| \tag{7.1}$$

The mean absolute percentage error (MAPE) is the percentage of the mean of the total error. On an excellent flood forecast, the MAPE should be smaller. It is written as:

$$\mathrm{MAPE} = \frac{1}{n}\sum_{i=1}^{n}\left|\frac{e_i}{y_i}\right| \times 100 \tag{7.2}$$

The root mean squared error (RMSE) is the square root of the mean of the squared deviation of the forecasted flood water level value. On an excellent flood forecast, the RMSE should be smaller. It is written as:

$$\text{RMSE} = \sqrt{\frac{1}{n}\sum_{i=1}^{n} e_i^2} \tag{7.3}$$

R^2 is the coefficient of determination and goodness of fit. With an excellent flood forecast, the R^2 should be larger.

$$R^2 = 1 - \frac{\text{sum squared regression (SSR)}}{\text{total sum of squares (SST)}} \tag{7.4}$$

The NerveNet LoRa data transmission performance is evaluated by calculating the packet delivery ratio (PDR) of LoRa packets.

$$\text{PDR} = \frac{\text{number of packets received}}{\text{number of packets sent}} \tag{7.5}$$

7.4 Results and Discussions

Table 7.3 compares the water level forecasting performance of the aforementioned five AI model types on the testing dataset. Theoretically, the deep learning methods outperform the conventional machine learning methods when the big data comes into its input. This is consistent with the result, where the LSTM and GRU models have a lower value of MAE, MAPE, and RMSE than the random forest and SVM models. This indicates that the deep learning models have a lower deviation of the forecasted results from the ground truth and a lower error percentage. A higher R^2 value indicates a more excellent time series forecasting performance from the deep learning models.

From the table, it can be observed that the LSTM model has more excellent performance than the GRU model, since it has lower MAE, MAPE, RMSE, and

Table 7.3 Benchmarking performance for prediction

AI model	MAE	RMSE	MAPE	R^2
Random forest	0.0656	0.078	0.0972	0.7807
SVM	0.0541	0.0632	0.0763	0.8562
GRU	0.0138	0.0154	0.0217	0.9915
LSTM (Keras)	**0.0088**	**0.0092**	**0.0126**	**0.997**
LSTM (OpenVINO)	0.0593	0.0907	0.0899	0.704

higher R^2. This finding is consistent with the findings in [11], where the LSTM model performs better than the GRU model in the case of short text processing and large-size datasets. In this paper, there is a huge amount of rainfall and water level dataset where both types of variables are short integers. They act as the inputs to the LSTM and GRU models. Therefore, it can be seen that the LSTM is more appropriate than the GRU models in these scenarios.

All in all, the LSTM has the best performance in the AI water level forecasting, since it has the lowest MAE, MAPE, and RMSE while the highest R^2 among all the proposed AI models. Therefore, LSTM is chosen as the AI water level forecasting model. Specifically, OpenVINO is used to convert the .h5 model to .xml and .bin format. It can be seen that there is a performance degradation of the converted model in all aspects.

Figure 7.1a displays the prediction versus ground truth for test dataset by using LSTM variations. As expected, the prediction using Keras model is close to the actual values. To reveal more insights, Fig. 7.1b compares the inference time between these two LSTM models. It can be seen that the LSTM (OpenVINO) is 28× slower than the Keras version. The reason is that the Keras model was using the Intel® Xeon® CPU at 2.20Ghz provided by the Google Colab. This hardware has more computational power than the NCS2, which consumes only around 1.5 W.

Figure 7.2 shows the actual deployment of LoRa nodes. For the LoRa parameters, we adopted spreading factor of 12, transmission power of 20 mW, and bandwidth of 500 kHz. Three NerveNet LoRa nodes serve as MQTT subscriber, whereas one NerveNet LoRa node acts as MQTT publisher. The publisher publishes the MQTT message at three different locations. At each location, a total of 11 LoRaMesh packets are transmitted. The quality of service (QoS) level is set to zero, which guarantees best-effort message delivery. In other words, the publisher only transmits each packet once, and LoRa message packets may be lost during the transmission process. Node 208 is located inside the building in such a way that nodes 203 and 204 can act as relay node. We implement subscriber and publisher nodes using Intel next unit computing (NUC) and Raspberry Pi 4, respectively. The latter is chosen due to its high portability and low cost, which is suitable for massive deployment of flood monitoring.

Figure 7.3 depicts the overall performance of NerveNet LoRaMesh. It can be observed from Fig. 7.3a that only extra hops are needed at location 3. This is reasonable since the distance between 204/208 and location 3 is at least 1200 m. In this case, node 203 which is closer to location 3 acts as relay node. For LoRaMesh packet to arrive at node 208, the packet initially sent by node 214 at location 3 is passed to 203, through 204 to 208. For other two locations, only one hop transmission is needed. This is because there are less obstacles, such as trees and buildings. The multi-hop transmission is affected by the received signal strength indicator (RSSI), as reported in Fig. 7.3b. All RSSI values are measured with respect to the publisher node 214, except the last two columns. Specially, 204 and 208 measurements are based on their relay nodes 203 and 204, respectively.

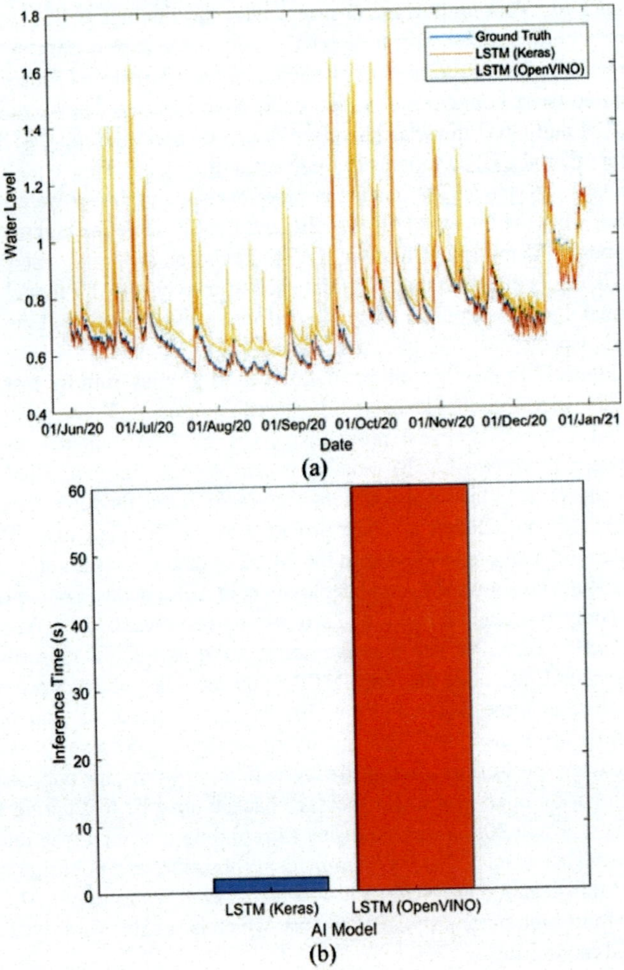

Fig. 7.1 LSTM performance benchmarking. (**a**) Prediction vs. ground truth. (**b**) Inference time

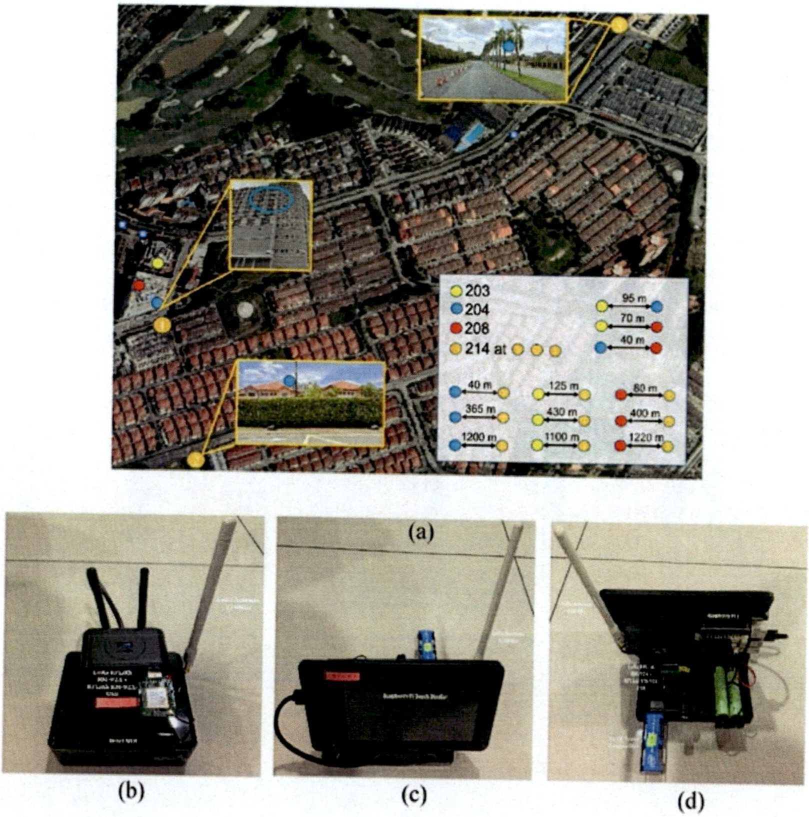

Fig. 7.2 System deployment. (**a**) The location of three subscriber nodes (203, 204, and 208) and one publisher node (214). (**b**) Subscriber node (Intel NUC). (**c**) Publisher node (front view). (**d**) Publisher node (rear view)

As shown in Fig. 7.3c, all LoRaMesh packets are received when the publisher transmits messages at locations 1 and 2. For location 3, 2 out of 11 packets are lost during the transmission for nodes 204 and 208. Specifically, when node 204 does not receive the packets from 203, it could not forward them to 208. Figure 7.3d compares the time on air. In LoRaMesh, time on air defines the elapsed time on air for a LoRaMesh packet between publisher and subscriber. As expected, the further the distance, the longer time needed to transmit the LoRaMesh packets.

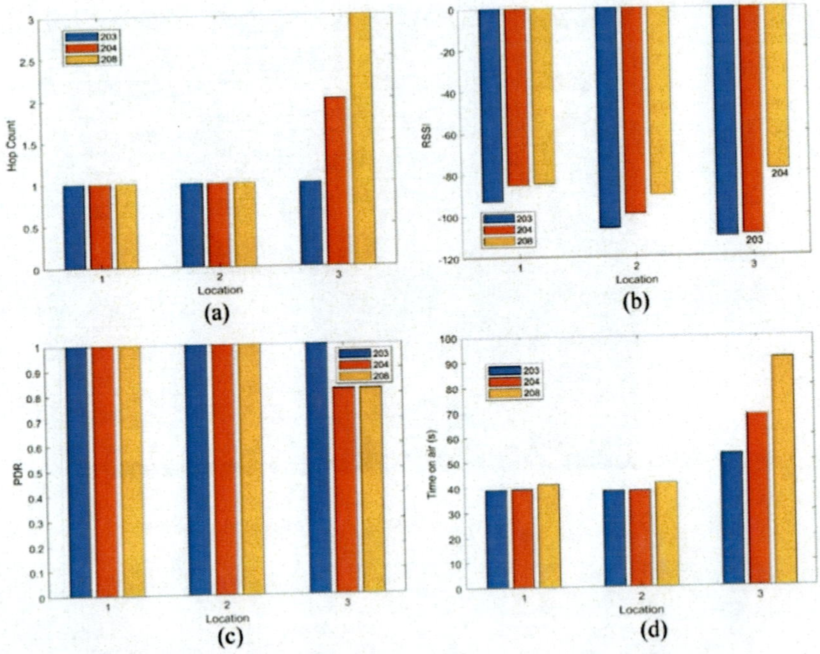

Fig. 7.3 Performance of NerveNet LoRaMesh. (**a**) Hop count. (**b**) RSSI. (**c**) PDR. (**d**) Time on air

7.5 Conclusion

In this paper, we have proposed an edge AI solution that forecasts flood water level and transmits the packet via LoRa mesh network. The AI model training and the testing dataset are obtained from Japan's organization. Hence, the AI results may not apply to the local area since the weather, season, humidity, and geographical condition of Malaysia are different from Japan. The local dataset can be requested from the local government to build an AI model that can fit the situation in Malaysia's local area so that a better understanding of the feasibility of the AI model in disaster detection in Malaysia.

Acknowledgments This work is the output of the ASEAN IVO (http://www.nict.go.jp/en/asean_ivo/index.html) project titled "Context-Aware Disaster Mitigation using Mobile Edge Computing and Wireless Mesh Network" and financially supported by NICT (http://www.nict.go.jp/en/index.html).

References

1. O.A. Kisi, A combined generalized regression neural network wavelet model for monthly streamflow prediction. KSCE J. Civ. Eng. **15**, 1469–1479 (2011)
2. Z.M. Yaseen et al., Artificial intelligence based models for stream-flow forecasting: 2000–2015. J. Hydrol. **530**, 829–844 (2015)
3. S. Hochreiter, J. Schmidhuber, Long short-term memory. Neural Comput. **9**(8), 1735–1780 (1997)
4. K. Cho, B.V. Merrienboer, D. Bahdanau, Y. Bengio. On the properties of neural machine translation: Encoder-decoder approaches, in Proceedings of SSST-8, Eighth Workshop on Syntax, Semantics and Structure in Statistical Translation. Association for Computational Linguistics, Doha, Qatar (2014), pp. 103–111
5. V. Mazzia, A. Khaliq, F. Salvetti, M. Chiaberge, Real-time apple detection system using embedded systems with hardware accelerators: An edge ai application. IEEE Access **8**, 9102–9114 (2020)
6. J.P. Queralta, T.N. Gia, H. Tenhunen, T. Westerlund. Edge-AI in LoRa-based health monitoring: Fall detection system with fog computing and LSTM recurrent neural networks, in Proceedings of 2019 42nd International Conference on Telecommunications and Signal Processing (TSP). IEEE, Budapest, Hungary (2019), pp. 601–604
7. M. Inoue, Y. Owada, NerveNet architecture and its pilot test in Shirahama for resilient social infrastructure. IEICE Trans. Commun. **100**(9), 1526–1537 (2017)
8. Ministry of Land, Infrastructure, Transport, and Tourism in Japan (MLIT Japan). Hydrology and Water Quality Database, http://www1.river.go.jp/. Last accessed 30 July 2022
9. N. Kimura et al., Convolutional neural network coupled with a transfer-learning approach for time-series flood predictions. Water **12**(96) (2020)
10. OpenVINO Toolkit. https://software.intel.com/enus/openvinotoolkit. Last accessed 01 July 2022
11. S. Yang, X. Yu, Y. Zhou. Lstm and gru neural network performance comparison study: Taking yelp review dataset as an example, in Proceedings of 2020 International Workshop on Electronic Communication and Artificial Intelligence (IWECAI). IEEE, Shanghai, China (2020), pp. 98–101

Chapter 8 Lightweight Certificateless Signature Scheme for Resource-Constrained IoT Environment

Chenghe Dong, Jianhong Zhang, and Lidong Han

Abstract With the widespread application of the Internet of things (IoT), the resource-constrained nature of IoT devices makes the security issues increasingly emerge. Certificateless cryptography has received significant attention due to the avoidance of complex certificate management and the removal of key escrow issues. A large number of certificateless signature schemes have been proposed; however, most of them have been shown to be insecure against public-key replacement attacks or malicious-but-passive KGC attacks. It is a challenge how to construct a secure and light certificateless signature suitable for the IoT yet. Recently, Xiang et al. proposed a lightweight certificateless signature scheme for the IoT which is claimed to be secure. However, by analyzing it, we find that it is vulnerable to public key replacement attacks. To overcome the above issue, we construct a secure and lightweight certificateless signature scheme in this work. It not only proves to be secure against two types of adversaries but also avoids time-consuming pairing. Finally, compared with recent several CLS signature schemes, evaluation results show that our scheme is comparable to the other schemes in terms of overall performance and security.

Keywords Certificateless signature · Internet of things (IoT) · Public key replacement attacks · Malicious-but-passive KGC attacks

8.1 Introduction

The Internet of Things (IoT) refers to the collective network of connected devices embedded sensors and the technologies that facilitate communication between them. Due to the advent of cheap computer chips and high-bandwidth telecommunications,

---

C. Dong · J. Zhang (✉)
North China University of Technology, Beijing, China
e-mail: chenghe_d@163.com; zjhncut@163.com

L. Han
Hangzhou Normal University, Hangzhou, China

C.-C. Wang, A. Nallanathan (eds.), *6th International Conference on Signal Processing and Information Communications*, Signals and Communication Technology, https://doi.org/10.1007/978-3-031-43781-6_8

we now have billions of constrained devices are now connected to the Internet of Things. However, limited computing power and storage capacity make the security of resource-constrained devices a huge challenge. Therefore, designing lightweight encryption with low energy consumption is more suitable for the security of resource-constrained devices.

With the advancement of information techniques, digital signature has become an indispensable component as it not only provides data integrity and non-repudiation of data resources but also verifies the authenticity of transactions. However, conventional encryption systems are no longer suitable to be implemented on resource-constrained IoT devices. In traditional digital signatures, public/private key generation relies on a trusted third-party certificate authority (CA). Before using the public key, its validity must be verified to guarantee the relationship between the public key and the certificate, which brings a heavy computation burden to the user. To overcome the complex certificate management, Shamir [1] introduced ID-based cryptography (IBC) in 1984, where the user's identity information directly acts as a public key and does not require to be authenticated by a trusted authority. However, the user's private key must be generated by a trusted third party called key generation center (KGC), which makes the key escrow problem inevitable.

To overcome these defects, Al-Riyami and Patterson [2] presented a certificateless public key cryptosystem (CL-PKC), in which not only the key escrow problem of IBC is solved but also the certificate management issue is simplified. The user's private key in CL-PKC is cooperatively generated by the KGC and the user. Thus, a secure certificateless signature must be able to resist attacks from a malicious user (Type I attack) and a malicious but passive KGC (Type II attack).

Since the introduction of CL-PKC, a large number of certificatelss signature (CLS) schemes have been proposed. Although some schemes are claimed to be secure, they are shown to be insecure against Type I or Type II attacks. Thus, it is still a significant challenge to design a secure and efficient CLS. In 2018, Jia et al. [3] proposed a novel CLS scheme and claimed that their scheme is secure and proven to secure against two types of attacks. However, Xiang et al. [4] pointed out Jia et al.'s CLS scheme to be insecure against malicious-but-passive KGC attacks and gave an improved CLS scheme to overcome their attacks. Then they show that the improved CLS scheme is secure against Type I and Type II attacks. Since time-consuming pairing operator is avoided, their improved scheme is an excellent lightweight CLS scheme. It is well suited for resource-constrained IoT devices.

In this work, we first analyze Xiang et al.'s improved CLS scheme and show that their scheme is still insecure. Namely, their scheme suffers from the Type I attack, and a malicious user can forge a signature without the partial key issued by KGC. Finally, we proposed a secure lightweight CLS scheme. Our main contributions in this work are summarized as follows:

- We analyze the security of Xiang et al.'s CLS scheme and show that their scheme is insecure. It can't resist the public key replace attack from a malicious user.

- To construct a secure CLS scheme, we put forth a novel CLS scheme in the random oracle model, which is proved to be secure against Type I and Type II attacks.
- Since the novel CLS scheme does not involve expensive pairing operators, it has higher computational performance. Compared with the recent several CLS schemes, the results show that our scheme is comparable in terms of computational cost and communication overhead.

8.2 Related Work

Due to its prominent advantages, certificateless signature has attracted significant attention from various scholars, since it was proposed and instantly became a research hotpot. Al-Riyami et al. [2] first proposed CL-PKC and the definition of its security model. Subsequently, Huang et al. [5] pointed out that there exists public key replacement attacks (i.e., Type I attacks) in [2]. In 2004, Yum and Lee [6] gave the general structure of designing CLS. Later, the structure proved vulnerable to Type I attacks in [7]. They then presented a modified scheme, and the security analysis was carried out in a simplified security model. Due to the popular application of bilinear pairings in cryptography, Li et al. [8] proposed the first CLS scheme based on bilinear pairings. In [9], Yap et al. presented another pairing-based CLS scheme. However, Park [10] found that Yap et al.'s scheme was unsafe.

Liu et al. [11] first introduced the CLS scheme in the standard model. Then, an improved CLS scheme without ROM was presented by Yu et al. [12] with better computational efficiency and shorter system parameters. In 2014, Yuan and Wang [13] explained that [12] was found to be insecure against both kinds of adversaries and then gave a modified CLS scheme. However, the above schemes are based on expensive pairing operations, which are inefficient.

He et al. [14] proposed the first CLS scheme with no pairings, which was proven to be secure in the ROM. Later, [15, 16] found that their scheme cannot resist the Type II attack. In 2018, Jia et al. [3] proposed an efficient and secure CLS signature scheme for IoT deployment. Whereafter, Du et al. [17] found that the proposal [3] is insecure for Type I attacks. Later, Thumbur et al. [18] proposed a new CLS scheme, but Xu et al. [19] soon found that it is as easy to signature forgery attack as [3]. Recently, Xiang et al. [4] pointed out that Jia et al.'s scheme is still forgeable under the Type II attacks and gave an improved scheme, which is higher efficient since no pairing operators are involved. Unfortunately, it is also insecure, and we will discuss it in the following part. Thus, it is a challenge to design a lightweight and secure CLS scheme that is suitable for the resource-limited IoT.

8.3 Related Work

8.3.1 Complexity Assumption

Elliptic Curve Cryptography (ECC) An elliptic curve is a set of defined by $y^2 = x^3 + ax + b \bmod p$ satisfying $4a^3 + 27b^2 \neq 0 \bmod p$. The non-singular elliptic curve defined on the finite field F_p is denoted as $E(F_p)$, where p is a large prime number and $a, b \in F_p$. Otherwise, an infinite point O is needed. Let all points on an elliptic curve with infinite point O form an additive cyclic group G.

$$G = \{(x, y) : y^2 = x^3 + ax + b, 4a^3 + 27b^2 \neq 0\} \cup O$$

Given $nP = O$, $P \in G$, where n is the order of G and P is a generator of G. $kP = P + \ldots + P$ (k *times*) denotes the scalar multiplication of group G, where $k \in Z_n^*$.

Elliptic Curve Discrete Logarithm Problem (ECDLP) Given a q-order additive cyclic group G and $P \in G$, where P is a generator of G. In the probabilistic polynomial time (PPT), for $Q = kP$, where Q is a point. It is not feasible to calculate $k \in Z_q^*$ from Q and P with non-negligible probability.

8.3.2 Certificateless Signature

In a CLS scheme, it involves three participants: the key generation center KGC, the signer, and the verifier, and consists of seven polynomial algorithms which are described as follows:

1. **Setup:** Inputs a security parameter λ, the KGC outputs the system public parameter PP, and the system master key msk. PP is used as an implicit input to the following algorithm.
2. **Extract Partial Private Key (PPK):** Inputs the master key msk, and the user's identity ID, it outputs the user's partial private key d which is returned to the user via a secure channel.
3. **Set Secret Value:** The user executes this algorithm to set his secret value x_{ID}.
4. **Set Private Key:** The user executes this algorithm obtains his full private key SK_{ID}, on inputs (ID, x_{ID}, d).
5. **Set Public Key:** The user executes this algorithm to obtain his public key PK_{ID}, on inputs (d, x).
6. **Sign:** The user executes this algorithm to produce a signature σ on a message m, on inputs (ID, SK_{ID}).
7. **Verify:** Receiving the user's (ID, σ, PK_{ID}, m), the verifier according to the verification results return 1/0.

8.3.3 Security Model

Al-Riyami et al. [2] gave the first formal security model of CLS, which involves two types of adversaries: Type I adversary and Type II adversary. The Type I adversary A_1 is a malicious user, who can replace public key without knowing the system secret key and partial private key. The Type II adversary A_2 is a malicious-but-passive KGC, who knows the system secret key, but cannot replace the public key of the target user. In the following, we describe the two types of adversaries using **Game1** and **Game2**.

Game1 This game is performed by an interactive game between an attacker A_1 and a challenger C_1. C_1 maintains a user list L_u and hash list L_h. This game contains three phases and is described in detail as follows.

- **Init**: On inputting the security parameter λ, C_1 executes *Setup* algorithm to generate the system parameter *PP* and the system master key *msk*. C_1 sends *PP* to A_1 and saves *msk* secretly.
- **Queries**: A_1 adaptively issues the following queries to C_1 in a polynomial time.
  - **CreateUser (ID)**: A_1 submits an identity *ID*, C_1 first searches L_u to check if *ID* has already been created. If it has, C_1 returns PK_{ID}. Otherwise, C_1 executes the following algorithms **Extract PPK**, **Set Secret Value**, **Set Private Key**, **Set Public Key** and obtains $(d_{ID}, x_{ID}, PK_{ID})$. Then C_1 adds the tuple $(ID, d_{ID}, x_{ID}, PK_{ID})$ in L_u and sends PK_{ID} to A_1. In general, we assume that *CreateUser* always precedes other related oracles.
  - **Replace-Public-Key Oracle (ID, PK'_{ID})**: A_1 submits the public key PK'_{ID}, which is update to L_u with the tuple $(ID, \perp, \perp, PK'_{ID})$ by C_1. If A_1 cannot provide the PK'_{ID} corresponding secret value x'_{ID}, $x'_{ID} = \perp$.
  - **Extract-Secret-Value Oracle (ID)**: Inputs an identity *ID*, C_1 checks L_u, and returns the secret value x_{ID} to A_1.
  - **ExtractPPK Oracle (ID, PK_{ID})**: Inputs an identity *ID* and its public key PK_{ID}, C_1 searches *ID* in L_u, and the corresponding partial private key d_{ID} is returned to A_1 if it exists. Otherwise, $\perp$ is returned.
  - **Sign Oracle (ID,m)**: Inputs a message *m* with *ID*, C_1 executes **Sign** algorithm, and returns the signature σ such that $Verify(ID,m,\sigma,PP,PK_{ID}) = 1$, where the latest public key PK_{ID} will be stored in L_u.
- **Forgery:** The attacker A_1 outputs a forged signature σ^* on message m^* for the challenged identity ID^*.

  1. Verify $(ID^*, m^*, \sigma^*, PP, PK^*_{ID}) = 1$.
  2. A_1 cannot ask for the signature of m^* under ID^*.
  3. A_1 cannot issue **ExtractPPK Oracle** query with ID^*.

The probability of A_1 winning the game is defined as:

$$\mathrm{Adv}_{A1}(\lambda) = \left|\left[\mathrm{Verify}\left(ID^*, m^*, \sigma^*, PP, PK_{ID}^*\right) = 1\right] - 1/2\right|$$

Game2 This game is played by the following interactive game between an attacker A_2 and a challenger C_2. Similar to Game1, Game2 also involves three phases.

- **Init:** C_2 does the same as C_1 and then returns both *PP* and *msk* to A_2.
- **Queries:** A_2 adaptively submits **CreateUser**, **Replace-Public-Key Oracle**, Extract-**Secret-Value Oracle**, and **Sign Oracle** same as that in **Game1**, and C_2 returns corresponding response. Note that **ExtractPPK Oracle** doesn't need to be queried since A_2 can compute partial private keys for itself.
- **Forgery:** The adversary A_2 outputs a forged signature σ^* on the message m^* with (ID^*, PK_{ID}^*). A_2 wins the game if the following conditions are satisfied:

  1. Verify $(ID^*, m^*, \sigma^*, PP, PK_{ID}^*) = 1$.
  2. A_2 never issues **Sign Oracle** with (ID^*, m^*).
  3. A_2 cannot conduct **Extract-Secret-Value** with ID^*.
  4. A_2 never issues **Replace-Public-Key** with PK_{ID}^*.

The probability of A_2 winning the game is expressed as:

$$\mathrm{Adv}_{A2}(\lambda) = \left|\left[\mathrm{Verify}\left(ID^*, m^*, \sigma^*, PP, PK_{ID}^*\right) = 1\right] - 1/2\right|$$

8.4 Reviews of Xiang et al.'s Scheme

We briefly review Xiang et al.'s [19] CLS scheme here. Their scheme is given as follows:

- **Setup**: KGC takes a security parameter λ to produce the parameters (q, F_p, G, P), where G is an q-order additive group with generator P over the finite field F_p. And KGC randomly chooses $s \in Z_q^*$ and computes public key $P_{pub} = sP$. Select three secure hash functions H_i: $\{0, 1\}^* \to Z_q^*$ $(i = 1, 2, 3)$. Finally, KGC publishes system parameters $PP = \{q, G, P, P_{pub}, H_1, H_2, H_3\}$, and master secret key $msk = s$ is kept secretly.
- **Extract-PPK**: Given an user's identity ID, the KGC randomly picks $r \in Z_q^*$ to compute $R = rP$, $h_1 = H_1(ID, R, P_{pub})$ and $d = (r + h_1 s) \bmod q$. Then the partial private key (R, d) is sent to the user via a secure channel. Use the equation $dP = R + h_1 P_{pub} \bmod q$ to verify its validity.

- **Set-Secret-Value**: The user randomly chooses a secret value $x \in Z_q^*$ to compute $X = xP$.
- **Set-Public-Key**: The user sets $PK = (R,X)$.
- **Set-Private-Key**: The user sets $SK = (d,x)$.
- **Sign:** For the signed message m, the user with identity ID randomly picks $t \in Z_q^*$ to compute $T = tP$ and $h_2 = H_2(ID,T,PK)$ and computes $\tau = x^{-1}(h_2 t + h_3 d) \bmod q$, where $h_3 = H_3(ID,m,T,PK,P_{pub})$. Finally, the signature is $\sigma = (T, \tau)$.
- **Verify**: For a message-signature tuple (m,σ), the verifier first calculates $h_1 = H_1(ID,R,P_{pub})$, $h_2 = H_2(ID,T,PK)$, and $h_3 = H_3(ID,m,T,PK,P_{pub})$ and then it checks

$$\tau X = h_2 T + h_3 \left(R + h_1 P_{pub}\right). \tag{8.1}$$

If yes, the verifier output "1," else it outputs "0."

8.5 Attacks on Xiang et al.'s CLS Scheme

Recently, Xiang et al. [4] proposed an improved CLS scheme, which was secure against Type I and Type II attacks they claimed. Unfortunately, in the following, we will show that their scheme is insecure; namely, it suffers from public-key-replacement attacks launched by a malicious user. The concrete attack is given as follows.

Attack Let a malicious user act as an adversary A, who can forge a CLS signature on an arbitrary message without the partial private key. The detailed forgery is shown as follows:

1. Let ID^* and m^* denote the identity of the attacked target's identity and a forged message, respectively.
2. Firstly, we randomly choose $b,c \in Z_q^*$ to compute $X' = bP_{pub}$ and $R' = cP_{pub}$. And we set the public key $PK^* = (X', R')$.
3. Next A chooses a random number $a \in Z_q^*$ to compute $T' = aP_{pub}$, $h'_2 = H_2(ID^*,T', PK^*)$, $h'_3 = H_3(ID^*,m^*,T', PK^*,P_{pub})$.
4. It computes $\tau' = b^{-1}(a \cdot h'_2 + h'_3(c + h_1)) \bmod q$.
5. At last, the forged signature under public key PK^* and identity ID^* is $\sigma' = (T', \tau')$.

Next, we prove that the forged signature σ' can pass the verification equation since

$$\begin{aligned}
\tau' X' &= b^{-1}\left(a \cdot h'_2 + h'_3(c + h_1)\right)\left(bP_{pub}\right) \\
&= \left(a \cdot h'_2 + h_3(c + h_1)\right)P_{pub} \\
&= ah'_2 \cdot P_{pub} + h'_3\left(cP_{pub} + h_1 P_{pub}\right) \\
&= h'_2 T' + h'_3\left(R' + h_1 P_{pub}\right)
\end{aligned}$$

Obviously, our attack is valid, and an adversary who replaces the user's public key can forge the signature on an arbitrary message without the partial private key. Thus, Xiang et al.'s scheme cannot resist Type I attack.

8.6 Our Scheme

8.6.1 Our Construction

In this part, we propose a lightweight and secure CLS scheme. The following is a detailed description of the proposed scheme.

- **Setup:** KGC inputs a security parameter λ and outputs the parameters (q,F_p,G,P), where G is a q-order additive group of with generator P over the finite field F_p. And then KGC randomly chooses $s \in Z_q^*$ as master secret key msk and calculates public key $P_{pub} = $ sP. In addition, KGC sets three hash functions: H_i: $\{0, 1\}^* \rightarrow Z_q^* (i = 1,2,3)$. Finally, KGC publishes system public parameters $PP = \{q,G,P, P_{pub},H_1,H_2,H_3\}$.
- **Set-Secret-Value**: The user randomly sets a number $x \in Z_q^*$ as its secret value and computes $X = xP$.
- **Extract-PPK**: The user submits (ID,X) to the KGC, and the KGC randomly chooses $r \in Z_q^*$ to compute $R = rP$, $h_1 = H_1(ID,X,R,P_{pub})$ and $d = (r + h_1 s) \bmod q$. And partial private key (R,d) is returned to the user via a secure channel. The user can verify its validity by the equation $dP = R + h_1 P_{pub}$.
- **Set-Public-Key:** The user's public key is set as $PK = (R,X)$.
- **Set-Private-Key:** The private key of the user is set as $SK = (d,x)$.
- **Sign:** Given (PP,ID,SK,m), the user randomly picks $t \in Z_q^*$ to calculate $T = tX$, $h_2 = H_2(ID,T,R,X)$ and $h_3 = H_3(ID,m,T,R,X,P_{pub})$ and calculates $\tau = (xh_2 t + h_3 d + xh_1) \bmod q$. Then $\sigma = (T,\tau)$ as the signature of message m.
- **Verify:** After receiving a message-signature tuple (m,σ), the verifier calculates: $h_1 = H_1(ID,R,X,P_{pub})$, $h_2 = H_2(ID,T,R,X)$, $h_3 = H_3(ID,m,T,R,X,P_{pub})$. Then it verifies if

$$\tau P - h_1 X = h_2 T + h_3 \left(R + h_1 P_{pub}\right). \tag{8.2}$$

If yes, the verifier outputs "1"; otherwise, it outputs "0."
The following equations verify the validity of the scheme:

$$\begin{aligned} \boldsymbol{\tau P - h_1 X} &= \boldsymbol{(xh_2 t + h_3 d + xh_1)P - h_1 xP} \\ &= \boldsymbol{h_2 tX + h_3 dP} \\ &= \boldsymbol{h_2 T + h_3 \left(R + h_1 P_{pub}\right)} \end{aligned}$$

8.6.2 Security Proof

Based on the security model described in Sect. 8.3.3, the security proof of our scheme is provided as follows.

Theorem 1 In the random oracle, the proposed CLS scheme is secured against Type I and Type II adversaries. **Lemma1** and **Lemma2** show that the proposed scheme is unforgeable against Type I and Type II adversaries, respectively.

Lemma 1 In polynomial time, a challenger C_1 who can solve the ECDLP problem with advantage, there exists a Type I adversary A_1 who succeeds in Game1 with non-negligible probability ε with ROM.

$$\varepsilon \geq \left(1 - \frac{1}{q_{cu}}\right)^{q_{eq}} \frac{1}{q_{cu}}, \tag{8.3}$$

where q_{cu} and q_{ep} denote the number of **CreateUser** and **Extract-PPK** queries.

Proof The Type I adversary A_1 succeeds in **Game1** with probability ε. We can structure a challenger C_1 attempts to solve s in the ECDLP instance $(G,P,Q = sP)$, with the help of A_1.

H_1, H_2, and H_3 are simulated as random oracles. C_1 maintains four lists L_u, L_1, L_2, and L_3, where L_u is for created users, and L_1, L_2, and L_3 are for H_1, H_2, and H_3, respectively. And all lists are initialized to be empty. The game between A_1 and C_1 is conducted as follows:

Init Let $\tilde{ID}$ denotes the target identity. C_1 sets $PP = (G,P,P_{pub} = Q)$ as public parameters and sends to A_1.

Queries A_1 issues the following oracle queries, and then C_1 responds as follows:

- **CreateUser(ID):** A_1 submits an identity ID, and C_1 searches L_u to check if ID has already been created. If it has, C_1 returns PK. Otherwise, if $ID \neq \tilde{ID}$, C_1 randomly chooses $d,x,h_1 \in Z_q^*$ to compute $R = dP - h_1 P_{pub} \bmod q$, $X = xP$. If $ID = \tilde{ID}$, C_1 selects $r,x,h_1 \in Z_q^*$ to compute $R = rP$, $X = xP$, $d = \perp$.

  C_1 checks if there already exists an entry (ID,R,X,P_{pub},H_1) in the L_1 and C_1 aborts the game if $H_1(ID,R,X,P_{pub}) \neq h_1$. Otherwise, C_1 returns $PK = (R,X)$ and, respectively, adds (ID,R,X,d,x) and (ID,R,X,P_{pub},h_1) in the lists L_u and L_1.
- **Hash query**: C_1 answers A_1's hash queries as follows:
  - *H_1-query.* When A_1 queries H_1 with the input (ID,R,X,P_{pub}), C_1 checks if L_1 exists the entry (ID,R,X,P_{pub},h_1). If yes, C_1 returns h_1 to A_1. Otherwise, C_1 computes $h_1 = H_1(ID,R,X,P_{pub})$ and adds (h_1,ID,R,X,P_{pub}) in L_1. Finally, h_1 is returned to A_1.
  - *H_2-query.* When A_1 queries H_2 with the input (ID,T,R,X), C_1 first checks if there exists (h_2,ID,T,R,X) in L_2. h_2 is returned if it exists. Otherwise, C_1 randomly chooses $h_2 \in Z_q^*$ to set $H_2(ID,T,R,X) = h_2$. Finally, C_1 returns h_2 to A_1 and adds (ID,T,R,X,h_2) in L_2.

- H_3-*query*. When A_1 queries H_3 with the input (ID,m,T,R,X,P_{pub}), C_1 checks if there exists (ID,m,T,R,X,P_{pub},h_3) in L_3, h_3 is returned if yes. Otherwise, C_1 randomly chooses $h_3 \in Z_q^*$ to compute $h_3 = H_3(ID,m,T,R,X,P_{pub})$. Finally, C_1 returns h_3 to A_1 and adds (ID,m,T,R,X,P_{pub},h_3) in L_3.

- **Extract-PPK(ID)**. When A_1 issues a query with ID, if $ID = \tilde{ID}$, C_1 returns "$\perp$" and aborts the game. Otherwise, C_1 searches L_u and returns d to A_1.
- **Extract-Secret-Value (ID).** For this query, C_1 checks L_u and returns the secret value x to A_1 if (ID,R,X,d,x) exists in L_u. Otherwise, C_1 makes **CreateUser** with ID and gets back x. Note that if A_1 had issued **Replace-Public-Key** query with ID, then $\perp$ is returned.
- **Replace-Public-Key (ID,PK′).** If (ID,PK') is asked such a query by A_1, where $PK' = (R',X')$. C_1 looks up and updates the entry (ID,R',X',d,x), (ID,R',X',P_{pub}, h_1), (ID,T,R',X',h_2), $(ID,m,T,R',X',P_{pub},h_3)$ in L_u, L_1, L_2, and L_3, respectively. Here x is set as "$\perp$."
- **Sign (ID,m).** On receiving this query, C_1 first searches (ID,R,X,d,x), (ID,R,X, P_{pub},h_1), (ID,T,R,X,h_2), and (ID,m,T,R,X,P_{pub},h_3) in the lists L_u, L_1, L_2, and L_3, respectively.

  - If $ID \neq \tilde{ID}$ and $x \neq \perp$ (the public key has not been replaced), C_1 randomly selects $t,h_1,h_2,h_3 \in Z_q^*$, sets $h_1 = H_1(ID,R,X,P_{pub})$, $h_2 = H_2(ID,T,R,X)$, $h_3 = H_3(ID,m,T,R,X,P_{pub})$, and calculates $T = tX$ and $\tau = (xh_2t + h_3d + h_1x) \bmod q$. And adds (ID,R,X,d,x), (ID,R,X,P_{pub},h_1), (ID,T,R,X,h_2), and $(ID,m,T,R, X,P_{pub},h_3)$ to the list L_u, L_1, L_2, and L_3, respectively.
  - If $ID = \tilde{ID}$ or $x = \perp$, C_1 randomly selects $t,h_1,h_2,h_3 \in Z_q^*$ and computes $T = h_2^{-1}$

    $[\tau P - h_3(R + h_2P_{pub}) - h_1X]$. C_1 outputs $\sigma = (T,\tau)$ and adds (ID,R,X,d,x), (ID,R, X,P_{pub},h_1), (ID,T,R,X,h_2), and (ID,m,T,R,X,P_{pub},h_3) to L_u, L_1, L_2, and L_3, respectively.

  Finally, C_1 outputs $\sigma = (T,\tau)$ to the adversary.

Forgery The forged signature $\sigma^* = (T^*,\tau^*)$ is the output of (ID^*,m^*), and A_1 never submits ID^* to **Extract-PPK** and **Sign**. C_1 will abort the game if $ID^* \neq \tilde{ID}$. Otherwise, C_1 searches for the entries (ID^*,R^*,X^*,d^*,x^*), $(ID^*,R^*,X^*,P_{pub},h_1^*)$, $(ID^*, T^*,R^*,X^*,h_2^*)$, and $(ID^*,m^*,T^*,R^*,X^*,P_{pub},h_3^*)$ in L_u, L_1, L_2, and L_3. C_1 will abort the game if h_2^* or h_3^* is not in L_2 and L_3. Else, if σ^* is valid, the following equation holds:

$$\tau^*P^* - h_1^*X = h_2^*T^* + h_3^*\left(R^* + h_1^*P_{pub}\right) \tag{8.4}$$

According to the forking lemma [20], C_1 repeats the game twice in the same way, but each response to H_2 and H_3 hash queries is different. Then A_1 outputs two forged signatures (T^*,τ^1) and (T^*,τ^2) which satisfy:

$$\tau^1P^* - h_1^*X = h_2^1T^* + h_3^1\left(R^* + h_1^*P_{pub}\right) \tag{8.5}$$

$$\tau^2P^* - h_1^*X = h_2^2T^* + h_3^2\left(R^* + h_1^*P_{pub}\right) \tag{8.6}$$

Because T^*, R^*, P_{pub}, and X^* can be represented as the formats $T^* = t^*X^*$, $R^* = r^*P$, $P_{pub} = sP$, and $X^* = x^*P$, we have the following three relations according to Eqs. (8.4)–(8.6).

$$\tau^* = \left(x^*h_2^*t^* + h_3^*\left(r^* + h_1^*s\right) + h_1^*x^*\right) mod\ q \tag{8.7}$$

$$\tau^1 = \left(x^*h_2^1t^* + h_3^1\left(r^* + h_1^*s\right) + h_1^*x^*\right) mod\ q \tag{8.8}$$

$$\tau^2 = \left(x^*h_2^2t^* + h_3^2\left(r^* + h_1^*s\right) + h_1^*x^*\right) mod\ q \tag{8.9}$$

where t^*, x^*, and s are unknown for C_1. However, C_1 can solve the value s by using the following equations Eq. (8.7)–(8.9).

$$\left(\tau^* - \tau^1\right) = \left(h_2^* - h_2^1\right)x^*t^* + \left(h_3^* - h_3^1\right)\left(r^* + h_1^*s\right) \tag{8.10}$$

$$\left(\tau^* - \tau^2\right) = \left(h_2^* - h_2^2\right)x^*t^* + \left(h_3^* - h_3^2\right)\left(r^* + h_1^*s\right) \tag{8.11}$$

so

$$x^*t^* = \frac{\left(\tau^* - \tau^1\right) - \left(h_3^* - h_3^1\right)\left(r^* + h_1^*s\right)}{\left(h_2^* - h_2^1\right)} \tag{8.12}$$

Taking x^*t^* into Eq. (8.11), we can obtain the relation

$$D \cdot \left(r^* + h_1^*s\right) = 0 \bmod q \tag{8.13}$$

thus, $s = -\frac{r^*}{h_1^*} mod\ q$ can be extracted, where $D = [(\tau^* - \tau^2)(h_2^* - h_2^1) - (\tau^* - \tau^1)(h_2^* - h_2^2)] - [(h_2^* - h_2^1)(h_3^* - h_3^2) - (h_2^* - h_2^2)(h_3^* - h_3^1)]$.

It means that the ECDLP can be solved; namely, given $(G,P,Q = sP)$, we can obtain s. Obviously, it is in contradiction with the difficulty of solving the ECDLP.

In the following, we analyze the probability of solving the ECDLP in **Game1**. C_1 succeeds if:

- E_1: C_1 does not abort the game.
- E_2: σ^* is a valid forgery on (ID^*,m^*).
- E_3: When A_1 submit the forged signature (m^*,σ^*) in the *Forgery* phase, we have $ID^* = \tilde{ID}$.

C_1's advantage is

$$\varepsilon_1 \geq Pr[E_1 \wedge E_2 \wedge E_3] = Pr[E_1]Pr[E_2|E_1]Pr[E_3|E_1 \wedge E_2] \tag{8.14}$$

Firstly, there is

$$Pr[E_2|E_1] \geq \varepsilon \tag{8.15}$$

Then, the probability of E_1 happens is at least

$$Pr[E_1] \geq \left(1 - \frac{1}{q_{cu}}\right)^{q_{ep}} \tag{8.16}$$

In the forgery, when $ID^* = \tilde{ID}$, the probability is

$$Pr[E_3|E_1 \wedge E_2] \geq \frac{1}{q_{cu}} \tag{8.17}$$

From the above analysis and Eqs. (8.15)–(8.17), we have

$$\begin{aligned} \varepsilon_1 &\geq Pr[E_1 \wedge E_2 \wedge E_3] \\ &= Pr[E_1]Pr[E_2|E_1]Pr[E_3|E_1 \wedge E_2] \\ &\geq \left(1 - \frac{1}{q_{cu}}\right)^{q_{ep}} \frac{1}{q_{cu}} \varepsilon \end{aligned}$$

Lemma 2 In polynomial time, if there exists a Type II adversary A_2 who succeeds in **Game2** with non-negligible probability ε with ROM, then the ECDLP problem can be solved.

$$\varepsilon \geq \left(1 - \frac{1}{q_{cu}}\right)^{q_{ep}+q_{rp}} \frac{1}{q_{cu}} \varepsilon \tag{8.18}$$

The query times of **CreateUser**, **Extract-Secret-Value**, and **Replace-Public-Key** are represented by q_{cu}, q_{es}, and q_{rp}.

Proof A_2 succeeds in **Game2** with probability ε. Then we can structure a challenger C_2 attempts to solve x in the ECDLP instance $(G,P,Q = \alpha P)$, by using A_2.

C_2 does same as C_1. C_2 maintains four lists L_u, L_1, L_2, and L_3. The game between A_2 and C_2 is conducted as follows:

- **Init:** Let $\tilde{ID}$ denotes the target identity. C_2 randomly picks $s \in Z_q^*$, sets $P_{pub} = sP$, $PP = (G,P,P_{pub} = Q)$, and sends (PP,s) to A_2.
- **Queries:** A_2 issues the following oracle queries and then C_2 responds as follows:
  - **CreateUser(ID):** When A_2 issues a **CreateUser** query with ID, C_2 does same as C_1.
  - **Hash query:** C_2 answers just as **Game1**.
  - **Extract-Secret-Value (ID).** For this query, C_2 checks if $ID = \tilde{ID}$. If yes, C_2 returns "$\perp$" and aborts the game. Else, C_2 finds L_u and returns x to A_2.
  - **Replace-Public-Key (ID,PK′).** If A_2 issues such a query with (ID,PK'), where $PK' = (R',X')$. C_2 does same as C_1.
  - **Sign (ID,m).** C_2 answers this query just as C_1 does in **Game1**.

- **Forgery:** Finally, A_2 outputs the forged signature $\sigma^* = (T^*,\tau^*)$, under the constrains that A_2 never submits ID^* to **Extract-Secret-Value**, **Sign**, and **Replace-Public-Key** queries. C_2 will abort the game, if $ID^* \neq \tilde{ID}$. Otherwise, C_2 searches for the entries $(ID^*,R^*,X^*,P_{pub},h_1^*)$, (ID^*,T^*,R^*,X^*,h_2^*), $(ID^*,m^*,T^*,R^*,X^*,P_{pub},h_3^*)$, and (ID^*,R^*,X^*,d^*,x^*) in L_1, L_2, L_3, and L_u.

  According to forking lemma [20], C_2 replays A_2 with the same random tape and provides a distinct value of $h_2(h'_2)$ for H_2 oracle, and then A_2 outputs another forgery (T^*,τ') of message m^*, and it satisfies:

$$\tau' P^* - h_1^* X = h_2' T^* + h_3^* \left(R^* + h_1^* P_{pub}\right) \tag{8.19}$$

In addition, $T^* = t^*X$, $R^* = r^*P$, $P_{pub} = sP$ and $X = Q = \alpha P$. Thus, we have the following equations:

$$\tau^* = \left(\alpha h_2^* t^* + h_3^* \left(r^* + h_1^* s\right) + h_1^* \alpha\right) \tag{8.20}$$

$$\tau' = \left(\alpha h_2' t^* + h_3^* \left(r^* + h_1^* s\right) + h_1^* \alpha\right) \tag{8.21}$$

where t^* and α are unknown for C_2. By Eqs. (8.20)–(8.21), we can obtain $(\tau^* - \tau') = (h_2^* - h_2^{(1)})\alpha t^*$, namely, we have

$$\alpha t^* = \frac{\tau^* - \tau'}{h_2^* - h_2^{(1)}} \tag{8.22}$$

Taking αt^* into Eq. (8.20), we have

$$\alpha = \left(h_1^*\right)^{-1} \left[\tau^* - \frac{h_2^*(\tau^* - \tau')}{h_2^* - h_2^{(1)}} - h_3^* \left(r^* + h_1^* s\right)\right] \tag{8.23}$$

It means that given $(G, P, Q = \alpha P)$, the solution α of the ECDLP can be solved.

In the following, we analyze the probability that C_2 wins in Game2. C_2 succeeds if:

- E_1: When A_2 queries *Extract-Secret-Value* and *Replace-Public-Key* orcles, C_2 does not abort the game.
- E_2: σ^* is a valid forgery on (ID^*,m^*).
- E_3: For the forged signature (m^*,σ^*) submitted by A_2 in the *Forgery* phase, we have $ID^* = \tilde{ID}$.

Firstly, there is

$$Pr[E_2 | E_1] \geq \varepsilon \tag{8.24}$$

Then, the probability of E_1 happens is at least

$$Pr[E_1] \geq \left(1 - \frac{1}{q_{cu}}\right)^{q_{es}} \left(1 - \frac{1}{q_{cu}}\right)^{q_{rp}} \tag{8.25}$$

In the forgery, when $ID^* = \tilde{ID}$, the probability is

$$Pr[E_3|E_1 \wedge E_2] \geq \frac{1}{q_{cu}} \tag{8.26}$$

From the above analysis and Eqs. (8.24)–(8.26), we have.

8.7 Performance Evaluation

We will evaluate the performance of our scheme by comparing with other CLS schemes without pairing. To ensure the security level of a *1024-bits* RSA algorithm, we adopt an elliptic curve E: $y^2 = x^3 + ax + b \bmod p$, where $a,b \in F_p$, G is a q-order additive cyclic group over $E(F_p)$, and both p and q are prime numbers of length 160 bits. The fundamental operations are implemented using the public C/C++ cryptographic library MIRACL, and the computing times to run the basic operations shown in Table 8.1 are run in the environment: Intel Core i5 processor, 3.00GHz frequency, Windows 10 Professional operating system, and 8192 MB memory device.

In Table 8.2, we compare our scheme with several CLS schemes without pairing [3, 4, 19, 21, 22] in terms of computational cost, communication cost, and security. We give the computational complexity of signature and verification phases, since the computational cost is mainly derived from these two stages. From Fig. 8.1 and Table 8.2, it can be seen that the computational costs in [3, 21] are the least in the signing phase, requiring only $T_{sm} = 0.539$ ms. The scheme [4] requires $T_{ex} = 3.932$ ms, which takes the most time. Our scheme is the same computational costs as the schemes [3, 4] as $T_{sm} + T_{inv} = 0.728$ ms, which is slightly higher than [19, 21]. However, in the verification phase, the computational costs in our scheme require $4T_{sm} + 2T_a = 2.160$ ms, as do in [3, 4]. It is slightly superior to the schemes

Table 8.1 Notation and running time of several operations

Notation	Operation	Time(ms)
T_{sm}	A scalar multiplication on elliptic curve	0.539
T_a	A point addition on elliptic curve	0.002
T_m	A modular multiplication operation	0.001
T_{inv}	A modular inversion operation	0.189
T_h	A general hash operation	0.001
T_{ex}	A modular exponentiation operation	3.932

Table 8.2 Comparative of pairing-free CLS schemes

Scheme	Sign	Verify	Signature size	Type I	Type II
Jia [3]	$T_{sm} + T_{inv}$	$4T_{sm} + 2T_a$	$\lvert G\rvert + \lvert Z_q^*\rvert$	No	No
Xu [19]	T_{sm}	$4T_{sm} + 4T_a$	$\lvert G\rvert + \lvert Z_q^*\rvert$	Yes	Yes
Yeh [21]	T_{sm}	$4T_{sm} + 3T_a$	$\lvert G\rvert + \lvert Z_q^*\rvert$	No	No
Xiang [4]	$T_{sm} + T_{inv}$	$4T_{sm} + 2T_a$	$\lvert G\rvert + \lvert Z_q^*\rvert$	No	Yes
Wang [22]	T_{ex}	$4T_{ex}$	$\lvert G_1\rvert + \lvert Z_q^*\rvert$	Yes	Yes
Our	$T_{sm} + T_{inv}$	$4T_{sm} + 2T_a$	$\lvert G\rvert + \lvert Z_q^*\rvert$	Yes	Yes

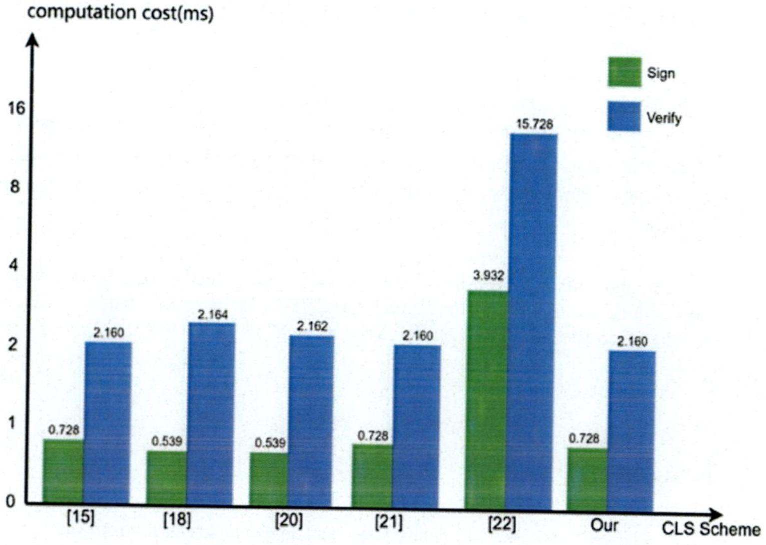

Fig. 8.1 Comparison of the computation costs

[19, 21] where computational costs are $4T_{sm} + 3T_a = 2.162$ ms and $4T_{sm} + 4T_a = 2.164$ ms, respectively. Meanwhile, all these schemes perform better than [22], and it needs $4T_{ex} = 15.728$ ms.

It is noted that the schemes [3, 4, 21] are insecure. They cannot resist the attack from Type I or Type II. However, our scheme can resist two types of attacks as [19, 22]. In addition, the size of the signature is also a critical factor which has an effect on communication overhead. As shown in Table 8.2, communication overhead in our scheme is the same as [4, 19, 21]. It is only $|G| + |Z_q^*| = 480 bits$, where $|G|$ and $|Z_q^*|$ denote the point size in the elliptic curve group G and the size of a group element in Z_q^*, respectively. However, the scheme [4] needs $|G_1| + |Z_q^*| = 1184 bits$ is more than twice the size of the above scheme, where $|G_1|$ denotes the point size of in the multiplicative cyclic group G_1.

In summary, our proposed scheme has more advantages over other several schemes in overall performance and security, since it is secure against Type I and Type II attacks with similar computational complexity and communication overhead.

8.8 Conclusion

The IoT devices are a type of resource-limited devices. Traditional signature scheme based on PKI is not suitable for the resource-limited device. In this paper, we analyze the security of a CLS scheme recently proposed by Xiang et al. [4] and point out that their scheme cannot resist Type I attacks. After analyzing the reasons to produce such attack, we propose a novel lightweight and secure CLS scheme. It can not only be proved to be secure against Type I and Type II adversaries but also avoid time-consuming pairing operators. Finally, compared with several recent CLS schemes, the results demonstrate that our scheme is the best in overall performance and security.

Acknowledgments This research was supported in part by the National Natural Science Foundation of China (no. 62172005), the Natural Science Foundation of Beijing (no. 4212019, M22002), and the Open Research Fund of Key Laboratory of Cryptography of Zhejiang Province (No. ZCL21014).

References

1. A. Shamir, Identity-based cryptosystems and signature schemes, in *Advances in Cryptology. CRYPTO 1984. Lecture Notes in Computer Science*, ed. by G.R. Blakley, D. Chaum, vol. 196, (Springer, Berlin, Heidelberg, 1985)
2. S.S. Al-Riyami, K.G. Paterson. Certificateless public key cryptography, in International Conference on the Theory and Application of Cryptology and Information Security (2003), pp. 452–473
3. X. Jia, D. He, Q. Liu, K.-K.R. Choo, An efficient provably-secure certificateless signature scheme for internet-of-things deployment. Ad Hoc Netw. **71**, 78–87 (2018)
4. D. Xiang, X. Li, J. Gao, X. Zhang, A secure and efficient certificateless signature scheme for internet of things. Ad Hoc Netw. **124** (2021)
5. X. Huang, W. Susilo, Y. Mu, F. Zhang. On the security of certificateless signature schemes from Asiacrypt 2003, in Cryptology and Network Security (2005), pp. 13–25
6. D.H. Yum, P.J. Lee, Generic construction of certificateless signature, in *Information Security and Privacy*, (Springer, Berlin, Heidelberg, 2004), pp. 200–211
7. B.C. Hu, D.S. Wong, Z. Zhang, X. Deng. Key replacement attack against a generic construction of certificateless signature, in Australasian Conference on Information Security and Privacy (2006), pp. 235–246
8. X. Li, K. Chen, L. Sun, Certificateless signature and proxy signature schemes from bilinear pairings. Lith. Math. J. **45**(1), 76–83 (2005)

9. W.S. Yap, S.H. Heng, B.M. Goi. An efficient certificateless signature scheme, in International Conference on Emerging Directions in Embedded and Ubiquitous Computing (2006), pp. 322–331
10. J.H. Park. An attack on the certificateless signature scheme from euc workshops 2006. Cryptology EPrint Archive (2006)
11. J.K. Liu, M.H. Au, W. Susilo. Self-generated-certificate public key cryptography and certificateless signature/encryption scheme in the standard model: Extended abstract, in Proceedings of the 2nd ACM Symposium on Information, Computer and Communications Security, Association for Computing Machinery (2007), pp. 273–283
12. Y. Yu, Y. Mu, G. Wang, Q. Xia, B. Yang, Improved certificateless signature scheme provably secure in the standard model. IET Inf. Secur. **6**(2), 102–110 (2012)
13. Y. Yuan, C. Wang, Certificateless signature scheme with security enhanced in the standard model. Inf. Process. Lett. **114**(9), 492–499 (2014)
14. D. He, J. Chen, R. Zhang, An efficient and provably secure certificateless signature scheme without bilinear pairings. Int. J. Commun. Syst. **25**(11), 1432–1442 (2011)
15. J.-L. Tsai, N.-W. Lo, T.-C. Wu, Weaknesses and improvements of an efficient certificateless signature scheme without using bilinear pairings. Int. J. Commun. Syst. **27**(7), 1083–1090 (2014)
16. M. Tian, L. Huang, Cryptanalysis of a certificateless signature scheme certificateless signature scheme without bilinear pairings. Int. J. Commun. Syst. **25**(11), 1432–1442 (2011)
17. H. Du, Q. Wen, S. Zhang, M. Gao, A new provably secure certificateless signature scheme for internet of things. Ad Hoc Netw. **100** (2020)
18. G. Thumbur, G.S. Rao, P.V. Reddy, N. Gayathri, D.R.K. Reddy, Efficient pairing-free certificateless signature scheme for secure communication in resource-constrained devices. IEEE Commun. Lett. **24**(8), 1641–1645 (2020)
19. Z. Xu, M. Luo, M.K. Khan, K.-K.R. Choo, D. He, Analysis and improvement of a certificateless signature scheme for resource-constrained scenarios. IEEE Commun. Lett. **25**(4), 1074–1078 (2021)
20. D. Pointcheval, J. Stern, Security arguments for digital signatures and blind signatures. J. Cryptol. **13**(3), 361–396 (2000)
21. K.-H. Yeh, C. Su, K.-K.R. Choo, W. Chiu, A novel certificateless signature scheme for smart objects in the internet-of-things. Sensors **17**(5) (2017)
22. L. Wang, K. Chen, Y. Long, H. Wang, An efficient pairing-free certificateless signature scheme for resource-limited systems. Sci. China Inf. Sci. **60**(11) (2017)

Chapter 9
NeSi: Netizen Simulator for Evaluating Internet Public Opinion Analysis System

Yan Yan, Mengjuan Fan, and Qingjia Luo

Abstract Clarifying potential public opinion information by analyzing comments is an important feature of modern Internet public opinion analysis systems. However, evaluating such systems by analyzing real netizens' comments requires significant human effort, which can be time-consuming and expensive. Netizen simulation is a cost-effective technique for evaluating Internet public opinion analysis systems. In this paper, we propose a conversational netizen simulator, called NeSi, for the automatic evaluation of such Internet public opinion analysis systems. Describing public opinion incidents, NeSi can automatically generate a sequence of comments based on the Opinion Leader-Follower model. The comments were varied and sentimental. Through a set of experiments that include both automatic and expert human evaluation, we show that NeSi generates comments that are not only related to public opinion but also comparable to human-generated utterances.

Keywords Internet Public Opinion Analysis System · Netizen simulator · Natural language processing

9.1 Introduction

The primary goal of an Internet public opinion analysis system is to predict trends in public opinion. To succeed, the system needs to have a clear understanding of netizens' comments and sentiments. Since Internet public opinion is often unclear and ambiguous, the system needs to actively collect comments from netizens. Clarifying the information needs of netizens has been shown to benefit both business

Y. Yan (✉) · Q. Luo
College of Information Engineering, Jiangmen Polytechnic, Jiangmen, China

Faculty of Data Science, City University of Macau, Macau, China
e-mail: d22092100108@cityu.mo

M. Fan
Jiangmen Internet Public Opinion Information Center, Jiangmen, China

C.-C. Wang, A. Nallanathan (eds.), *6th International Conference on Signal Processing and Information Communications*, Signals and Communication Technology, https://doi.org/10.1007/978-3-031-43781-6_9

and public services, providing a strong motivation for such an Internet public opinion analysis system.

However, the evaluation of the described system for Internet public opinion analysis is not straightforward. The main reasons are two aspects, expensive and time-consuming human-in-the-loop, and inconsistent human judges criteria.

Such research requires netizens to provide a large number of comments to the system in real time. A relatively simple solution is to conduct a corpus-based offline evaluation. However, this has limited the ability of the system to analyze dynamic public opinion, and the effect is not good in practical application. Moreover, such offline evaluation is still limited to the interactions of a single netizen, as the predefined news content is associated with corresponding comments and is unaware of the interactions of other netizens.

To address the shortcomings of both user-based and corpus-based evaluation methods, we propose netizen simulation. The simulated netizens aim to capture the behavior of real netizens, which is to generate multiple comments based on news content. This approach remains as scalable and inexpensive as other offline evaluation methods.

In this paper, we propose a conversational netizen simulator, called NeSi, which is a model capable of multi-turn interactions with a general Internet public opinion analysis system. Given some news content, NeSi generates comments in fluent and coherent natural language, making its responses comparable to real netizens. An example of NeSi evaluating an Internet public opinion analysis system is shown in Fig.9.1.

We formulate the main research question: "*How to model human-like comment behavior?*" To answer it, we base our proposed netizen simulator on a large-scale transformer-based language model, namely, BERT [1] and BART [2], ensuring the near-human quality of Natural Language Understanding (NLU) and Natural Language Generating (NLG). In addition, NeSi generates comments consistent with the initial news content to simulate the behavior of real netizens. With a specific training procedure, the language model can be guaranteed to be semantically controlled. We evaluate the feasibility of our approach through a series of detailed experiments that include automatic measurements as well as human judgments.

We describe the process of creating comment sequences, which involves creating a series of various expressions for the same news, each with a particular sentiment (positive, negative, neutral). The issues with question rewriting in conversational QA [3] and query suggestions in conversational information-seeking tasks [4] are similar to those in this task. Comment sequence generation produces a variety of comments according to various sentiment types, should the system be able to comprehend news content, as opposed to producing a single rewrite of the query. The sentiment type sequence of comments is preset by the evaluator himself.

The target sentiment type and news are inputted into the Opinion Leader module and Follower module to complete this process. Although this approach can provide a variety of sentimental expressions, not all of them are suitable because they do not deconstruct the language. As a result, we also look into how to assess the sentiment types of generated comments and exclude utterances that are regarded to be

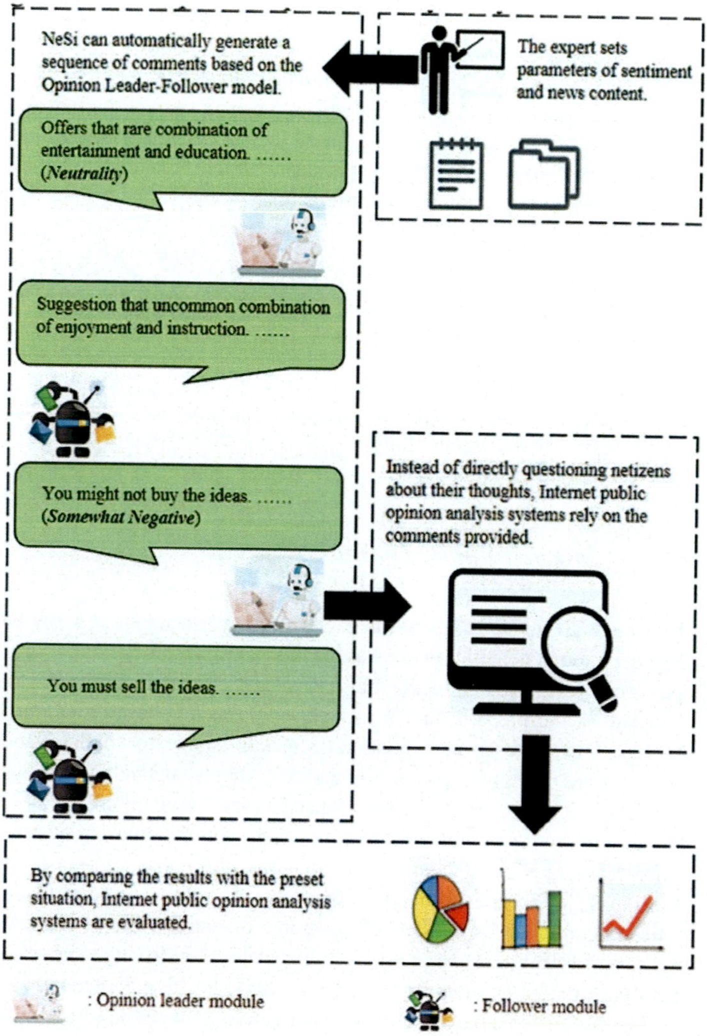

Fig. 9.1 An example of evaluation processes

linguistically inappropriate. Additionally, we carry out a contextual evaluation by replacing real-world human comment sequences with synthetic ones.

In summary, the novel contributions of this paper are as follows:

- We propose an opinion leader module built upon BERT and BART to generate sentimental comments. The module consists of a keyphrase extractor, content planner, and content filler. It extracts keyphrases from news content and generates comments according to preset sentimental expression requirements.

- We introduce the methods of estimating the reading difficulty and sentiment types. We use the CoLA dataset and SST-5 dataset to fine-tune the BERT-based classifier. The method is used to distinguish whether there are syntax errors in utterances and add sentiment labels to syntax correct utterances.
- We propose a follower module based on Part of Speech to rewrite sentimental comments. It is a flexible component, which can adjust the rewriting rules according to the evaluation requirements.

9.2 Related Work

This study lies in the intersection of Internet public opinion analysis and user simulation. So we refer to Internet governance and natural language processing（NLP）technology.

9.2.1 Internet Public Opinion Analysis

By March 2021, the number of netizens in the world stands at 5.169 billion, and the highest Internet penetration rate has reached 93.9 percent in North America. Over the past few years, social media platforms (e.g., Twitter, Facebook, TikTok) have reached a widespread diffusion as a personal and handy information channel. Most scholars have chosen user reviews as a data source to conduct studies from the perspectives of segmenting communication phases, analyzing communication characteristics, and communication influence factors. Based on life cycle theory, scholars divide the law of public opinion propagation on the Internet of emergencies into different stages (i.e., latent, fermentation, outbreak, remission, recurrent, and decline) [5]. Cao and Lu believed that Internet public opinion had three characteristics: sentimental expression of opinions, diversification of communication subjects, and self-interest of media [6]. Lin et al. established a model of influencing factors of public opinion propagation in public health emergencies [7].

The comments have recently been found to be a potential source of important data because they are public, simple to crawl, and can have their content examined using the right text/data mining tools. We carefully selected four current Internet public opinion analysis systems, namely, Istarshine,[1] Antfact,[2] MIDU,[3] and Izhonghong,[4] each of which should at least support the text sentiment analysis and set retrieval functionalities.

---

[1] https://www.istarshine.com/

[2] https://www.eefung.com/

[3] https://yqt.midu.com/

[4] http://www.izhonghong.com/

9.2.2 User Simulation in Natural Language Processing

In the past, agenda-based or model-based simulation has been widely used to train conversation state-tracking components of conversation agents using reinforcement learning technology. The highly interactive nature of conversational information access systems has also aroused people's interest in the use of user simulation in the field of information retrieval. Recently, Zhang and Balog proposed a general method to evaluate conversational recommendation systems through user simulation [8]. They use NLU, NLG, and response generation in an agenda-based simulator. However, building a humanoid simulator is still a difficult task [9]. To create a more realistic simulator for information-seeking dialogue, Salle et al. put great emphasis on such behavioral characteristics as patience and cooperation [10]. Sekulic et al. went further to enable analog users to ask clarifying questions in the hybrid initiative setting [11]. Sun et al. studied how to simulate task-oriented user satisfaction in a human-like manner [12].

In this work, we focus on improving human-likeness in terms of how netizens formulate their words for topic discussion. Since netizens tend to connect with like-minded people and express their opinions on current affairs on social network, Internet public opinions are naturally generated and social. Opinion leaders play a more and more important role in the process of opinion dissemination and information transmission under their influence and even affect the direction of decision-making [13]. We further use natural language to generate human-like comments.

9.3 Simulation Methodology

Most Internet public opinion analysis systems are still in an early stage of development, and many questions have not been well studied, including netizen behavior on social network. We carry out a netizen study to answer the following question: *What basic behaviors of netizens need to be monitored by the Internet public opinion analysis system?* To ensure that our observations are based on realistic netizen behavior, we need to understand how the Internet public opinion analysis system works. Based on the collected datasets, we find that Opinion Leader-Follower is the most frequent type of behavioral pattern. Although the Internet public opinion analysis system only pays attention to the comments of opinion leaders, the collective sentimental trend of netizens is also very valuable.

Most of the previous works in Internet public opinion analysis systems have used corpus-based approaches to evaluate the performance of the system in an offline setting. However, offline evaluation can only be performed at the static level, which does not accurately reflect the nature of such systems. Therefore, it is necessary to involve citizens in the evaluation process to correctly capture the nature of the Internet public opinion search task. While this approach accurately captures the performance of the system in real-world scenarios, it is tedious, expensive, and not

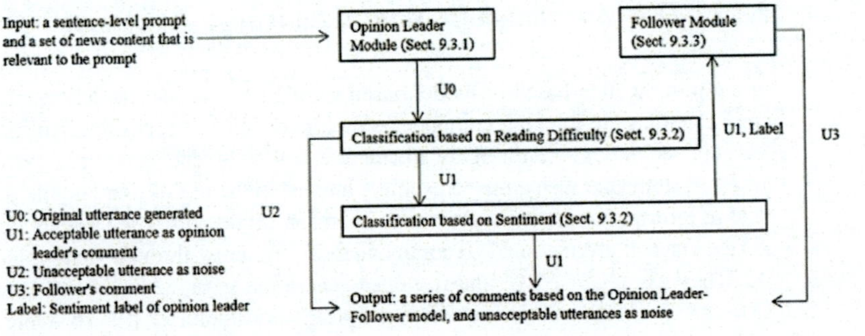

Fig. 9.2 Netizen simulator architecture

scalable. To reduce the evaluation of the Internet public opinion analysis system and still accurately capture the overall performance, we propose a simulated netizen method. Simulated netizens are designed to provide an alternative to real netizens, because it is easy to scale, cheap, fast, and consistent. The architecture we designed for this task is illustrated in Fig.9.2.

Next, we will discuss how to build an opinion leader module to generate comments based on the given news content (Sect. 9.3.1). To ensure that the comments achieve the desired purpose, we further study the methods of estimating the generated comment types (Sect. 9.3.2). For these comments, the follower module can make some opinions (Sect. 9.3.3). This is a control text generation task. So we refer to the method of the PAIR [14] and modify it according to the characteristics of the dialogue model between opinion leaders and followers.

9.3.1 *Opinion Leader Module*

The input of the opinion leader module includes (1) a sentence level prompt, such as a topical subject with a sentimental slant in a conversation, and (2) a set of news content that is relevant to the prompt. The simulator aims to generate utterance that contains multiple sentences with different sentiments as an argument, by reflecting the news content coherently. To be more specific, the opinion leader simulator has the following three steps to achieve the abovementioned design objective.

Step 1: Keywords Extraction. We combine long text with a series of news items from the same topic. Then we use KeyBERT [15] tool to extract keyphrases. KeyBERT is an end-to-end tool for keyword and keyphrase extraction. It can self-label unlabeled corpus, reduce manual work, and extract context keywords. This method randomly uses data from Wikipedia web pages and conducts training through two-way LSTM [16]. In the self-labeling stage, the bidirectional conversion encoder is used to extract the context features from the text, which is superior to the keyword tags obtained by some traditional methods.

Step 2: Content Planning. By training BERT, our content planners assign keyphrases to different sentences and predict their corresponding positions. We use bidirectional self-attentions for input encoding and apply causal self-attentions for keyphrases assignment and position prediction. There are three special tokens. [BOK] token signals the beginning of keyphrase assignment generation. [SEN] token is generated to represent the sentence boundary. [EOS] token signals the planning process terminates.

Step 3: Content Plan Addition. We use the seq2seq generation framework with BART fine-tuning, which includes a given content plan derived from a keyphrase. We use a content planning model to output keyphrase assignments for different sentences. We use post-processing steps to convert between different tokenizers. The keyphrases of each sentence specified are placed in the position corresponding to the prediction, and the empty slot is filled with [MASK] tags. The input of the encoder is composed of an input prompt, keyword assignment, and template concatenation. The decoder generates the output original utterance.

9.3.2 *Classification Based on Reading Difficulty and Sentiment*

To avoid generating opinion leader comments that are harder to understand or more complex than real comments, we include a classifier to estimate reading difficulty in this module. It classifies the utterance output from each opinion leader module as acceptable or unacceptable. Only the accepted utterance can be used as the input of the next classifier (i.e., the sentiment classifier). We use the CoLA dataset to fine-tune the BERT-based classifier. The dataset is annotated according to whether the language is linguistically acceptable or not. Three types of sentences are marked as unacceptable (i.e., morphological, syntactic, and semantic anomalies).

Since conversations between opinion leader and follower should also contain sentiment information, it is necessary to perform sentiment analysis on opinion leader. We include a classifier for sentiment analysis in our simulations. We use the SST-5 dataset to fine-tune the BERT-based classifier. SST-5 is a standard dataset for fine-grained sentiment classification, including five types of sentimental discourse (i.e., negative, somewhat negative, neutral, somewhat positive, and positive).

After passing through the two classifiers, the utterances generated by the opinion leader simulator are divided into unacceptable utterances and acceptable utterances with sentiment labels.

9.3.3 Follower Module

Follower module input consists of acceptable utterances and sentiment labels. The simulator aims to rewrite the utterances of the opinion leader according to the preset rules. First, we extract the Part-of-Speech tags from the acceptable utterances. Next, for acceptable utterances with sentiment labels, we choose and perform one of the following operations.

- If the label is neutral, the noun and adjective of the utterances are replaced with synonyms.
- If the label is somewhat negative or somewhat positive, the verb and adjective of the utterances are replaced with antonyms.
- If the label is negative or positive, half of the words in the utterances are randomly selected for synonym replacement.

Then the acceptable utterances and the rewritten sentences are combined into a pair of comments. Our investigation led us to the conclusion that more than half of the viewpoints that may be found on the Internet are merely noise. Real netizens often use meaningless utterances to avoid being monitored by the Internet public opinion analysis system. So unacceptable utterances can be added proportionally to comments as a kind of noise. In this way, the system can be evaluated with stricter standards.

9.4 Experiments

9.4.1 Experimental Data and Methods

We collected online data on five public opinion incidents from four Internet public opinion analysis systems (Sect. 9.2.1). We structured this data and further enriched it with another crowd-sourcing task to construct training data for the comment generation task. We only collect opinions related to the incidents. To extract comment behavior from recorded conversations, expert notes (by the two authors of the paper) follow three types of emotions at the communication level: positive, negative, and neutral.

We enrich the initial datasets by performing a crowd-sourcing task of discourse rewriting. We showed workers an interactive scene where an opinion leader and a follower talk about news content. The original annotated sequence is used as input to the simulation scenario, and each rewrite task is enabled. We invited ten annotators to rewrite each utterance according to our intent and obtained about 2.0 k rewritten comments.

We used the pretraining model tool provided by the Huggingface platform [17] to build NeSi. To test the generalization ability of NeSi, we use training data in two ways: single and hybrid. The single mode uses the topic dataset with the highest score as the training dataset. The hybrid mode is to combine five topic datasets and then randomly select a portion of them as the training dataset. The training process was conducted through fivefold cross-validation. The parameters to initialize the model are default parameters and batch size 10. We test the model following the comment generation procedure in Fig. 9.2.

9.4.2 Experimental Evaluation

We evaluate the performance of our designed NeSi using both automatic and expert human evaluation methods. NeSi mainly performs the task of automatic comment sequence generation.

Use NLP indicators for automatic evaluation. To verify the rationality of the NeSi component combination, we use ROUGE [18] and BLEU [19] for automatic evaluation. These two indicators are often used in machine translation, automatic summary, question and answer generation, and other fields. We tested the significance of the following pairs of methods: only using the opinion leader module (OOL) vs. opinion leader module with classifier (OL-C) and opinion leader module with classifier (OL-C) vs. NeSi.

To ensure the stability of the experimental results, we generated the comment sequence ten times. We calculate the metrics using micro-averaging and further average the score across each sequence. All results are shown in Table 9.1. We have noticed that only the opinion leader module (OOL) has a much lower performance than the other two methods. The opinion leader module with a classifier (OL-C) can improve performance in some cases. NeSi achieves optimal performance in almost all cases. The follower module does not degrade system performance. This indicates that the structure of NeSi is reasonable. The performance of the hybrid mode is better than the single mode, which shows the generalization ability of the hybrid model.

Use expert human evaluation. We investigate the end-to-end performance of our model using expert human evaluation. Experts assess whether the simulated utterances generated by NeSi are indistinguishable from the conversations formed by real netizens. We extracted 100 conversations from the recorded data and replaced the original conversation snippets with simulated utterances of the same comment type. We randomly generated a total of 200 dialogue segments. We presented ten experts with random snippets of both original and simulated conversations. They were asked to choose which of these conversation segments were the simulated utterances

Table 9.1 Automatic evaluation results

	Single				Hybrid			
Method	Rouge-1	Rouge-2	Rouge-L	BLEU	Rouge-1	Rouge-2	Rouge-L	BLEU
OOL	0.272	0.094	0.267	2.147	0.304	0.103	0.304	2.473
OL-C	0.495	0.133	0.493	5.183	0.519	0.151	0.493	7.839
NeSi	0.504	0.137	0.506	6.204	0.573	0.263	0.516	8.295

produced by NeSi. They are specifically directed toward, which is a comment sequence based on the Opinion Leader-Follower model. When it is difficult to distinguish, abstention is permitted. In addition, experts were asked to give a brief explanation of their choices. We find that the improvement of the opinion leader module and classifier is consistent with both automatic and expert manual evaluation. Notably, 32.5% of the pairs are distinguishable. This may be due to the simple rules of the follower module. In the future, higher-level intelligent rules can be used for the follower module as needed to improve the human-like nature of the simulated responses.

9.5 Conclusion

This work focuses on a specific netizen feature, discourse comments in the Opinion Leader-Follower model, to create a more human-like simulation of netizens. We have developed a tool for evaluating the Internet public opinion analysis system, which is composed of an opinion leader module, comment-type classifier, and follower module. The proposed methods have been compared using automatic evaluation and expert human evaluation.

Limitations Our netizen research focuses on type 1:1 leader-follower behavior. In practice, the simulator needs to model netizens with different organizational types, ranging from type 1: n to type m: n. We also note that our model does not take into account the differences in the comments of netizens on different social media platforms.

Future Work We see several ways to improve our work in the future. More human-like comment behavior will be incorporated into a larger framework of netizen simulation. There are other aspects of commenting behavior that needs to be considered, such as the knowledge and preferences of netizens, changing intentions, and consistency in the style of comment generation.

Acknowledgments The authors would like to thank the reviewers for their constructive feedback and all the experts who participated in this study. They would also like to thank Assist. Prof. Zuobin Ying of the Faculty of Data Science at the City University of Macau for his insightful comments. This research is supported by NSFC-FDCT under its Joint Scientific Research Project Fund（Grant No. 0051/2022/AFJ), China & Macau

References

1. Jacob Devlin, Ming-Wei Chang, Kenton Lee, Kristina Toutanova. BERT: Pre-TRAINING of deep bidirectional transformers for language understanding, in Proceedings of the 2019 Conference of the North American Chapter of the Association for Computational Linguistics (ACL) (2019), pp. 4171–4186

2. P. Lippe, P. Ren, H. Haned, B. Voorn, M. de Rijke. Diversifying task-oriented dialogue response generation with prototype guided paraphrasing. IEEE/ACMTrans. Audio Speech Lang. Process (2021). https://doi.org/10.48550/arXiv.2008.03391
3. Qian Liu, Bei Chen, Jian-Guang Lou, Bin Zhou, Dongmei Zhang. In complete utterance rewriting as semantic segmentation, in Proceedings of the 2020 Conference on Empirical Methods in Natural Language Processing (EMNLP) (2020), pp. 2846–2857
4. Corbin Rosset, Chenyan Xiong, Xia Song, Daniel Campos, Nick Craswell, Saurabh Tiwary, Paul Bennett. Leading conversational search by suggesting useful questions, in Proceedings of The Web Conference (WWW) (2020), pp. 1160–1170
5. R. Zeng, C. Wang, Q. Chen, Comparative research on the stages and models of network public opinion dissemination. J Inf **33**(5), 119–124 (2014)
6. W. Cao, H. Lu, The formation and guidance of public opinion in social media in the era of mobile internet-taking the WeChat dissemination of the "Shandong vaccine incident" as an example. Southeast Commun **06**, 56–58 (2016)
7. L. Wang, K. Wang, W. Jiang, The spread and evolution of public opinion on public health emergencies in social media: Taking the vaccine incident in 2018 as an example. Data Anal Knowl Discov **3**(04), 42–52 (2019)
8. Shuo Zhang, Krisztian Balog. Evaluating conversational recommender systems via user simulation, in Proceedings of the 26th ACM SIGKDD International Conference on Knowledge Discovery and Data Mining (KDD) (2020), pp. 1512–1520
9. Krisztian Balog. Conversational AI from an information retrieval perspective: Remaining challenges and a case for user simulation, in Proceedings of the 2nd International Conference on Design of Experimental Search & Information REtrieval Systems (DESIRES) (2021), pp. 80–90
10. Alexandre Salle, Shervin Malmasi, Oleg Rokhlenko, Eugene Agichtein. Studying the effectiveness of conversational search refinement through user simulation, in Proceedings of the 43rd European Conference on Information Retrieval (ECIR) (2021), pp. 587–602
11. Ivan Sekulic, Mohammad Aliannejadi, Fabio Crestani. Evaluating mixed-initiative conversational search systems via user simulation, in Proceedings of the 15th International Conference on Web Search and Data Mining (WSDM) (2022), pp. 888–896
12. Weiwei Sun, Shuo Zhang, Krisztian Balog, Zhaochun Ren, Pengjie Ren, Zhumin Chen, Maarten de Rijke. Simulating user satisfaction for the evaluation of task-oriented dialogue systems, in Proceedings of the 44th International ACM SIGIR Conference on Research and Development in Information Retrieval (SIGIR) (2021), pp. 2499–2506
13. A. Yl, B. Gl, K.A. Gang, Consensus reaching process in large-scale group decision making based on opinion leaders. Procedia Comput Sci **199**, 509–516 (2022)
14. Xinyu Hua, Lu Wang PAIR: Planning and iterative refinement in pre-trained transformers for long text generation, in Proceedings of the 2020 Conference on Empirical Methods in Natural Language Processing (EMNLP) (2020), pp. 781–793
15. Prafull Sharma, Yingbo Li. Self-supervised contextual keyword and keyphrase retrieval with self-labelling. https://maartengr.github.io/KeyBERT/
16. A. Graves, J. Schmidhuber, Framewise phoneme classification with bidirectional LSTM and other neural network architectures. Neural Netw. **18**(5–6), 602–610 (2005)
17. Thomas Wolf, Lysandre Debut, Victor Sanh, Julien Chaumond, Clement Delangue, Anthony Moi, Pierric Cistac, Tim Rault, Remi Louf, Morgan Funtowicz, Joe Davison, Sam Shleifer, Patrick von Platen, Clara Ma, Yacine Jernite, Julien Plu, Canwen Xu, Teven Le Scao, Sylvain Gugger, Mariama Drame, Quentin Lhoest, and Alexander Rush. Transformers: State-of-the-art natural language processing, in Proceedings of the 2020 Conference on Empirical Methods in Natural Language Processing (EMNLP) (2020), pp. 38–45

18. Chin-Yew Lin. ROUGE: A package for automatic evaluation of summaries, in Proceedings of the ACL workshop on Text Summarization Branches Out (2004), pp. 74–81
19. Kishore Papineni, Salim Roukos, Todd Ward, Wei-Jing Zhu. BLEU: A method for automatic evaluation of machine translation, in Proceedings of the 40th Annual Meeting of the Association for Computational Linguistics (ACL) (2002), pp. 311–318

Index

C.-C. Wang, A. Nallanathan (eds.), *6th International Conference on Signal Processing and Information Communications*, Signals and Communication Technology, https://doi.org/10.1007/978-3-031-43781-6